Robert Angus Smith

Memoir of JOHN DALTON And History of the Atomic Theory up to His Time

Elibron Classics
www.elibron.com

Elibron Classics series.

ISBN 1-4021-6437-8 (paperback)
ISBN 1-4021-1417-6 (hardcover)

This Elibron Classics Replica Edition is an unabridged facsimile of the edition published in 1856 by H. Bailliere, London.

MEMOIRS

OF THE

LITERARY AND PHILOSOPHICAL SOCIETY

OF

MANCHESTER.

MEMOIRS

OF THE

LITERARY

AND

PHILOSOPHICAL SOCIETY

OF

MANCHESTER.

Second Series.

VOLUME THIRTEENTH.

LONDON:
H. BAILLIERE, 219, REGENT STREET,
AND 290, BROADWAY, NEW YORK.
PARIS: J. B. BAILLIERE, LIBRAIRE, RUE HAUTEFEUILLE.
1856.

MANCHESTER:
PRINTED BY T. SOWLER AND SONS, ST. ANN'S SQUARE.

W. Morton. Sc. Manchester.

MEMOIR

OF

JOHN DALTON,

D.C.L., F.R.S.,
INSTIT. (ACAD. SC.) PARIS. SOCIUS.,
PRESIDENT OF THE LITERARY AND PHILOSOPHICAL SOCIETY OF MANCHESTER, ETC., ETC.;

AND

HISTORY

OF

THE ATOMIC THEORY

UP TO HIS TIME.

BY

ROBT. ANGUS SMITH, PH.D., F.C.S.

SEC. TO THE LIT. AND PHIL. SOC.

LONDON:
H. BAILLIERE, 219, REGENT STREET,
AND 290, BROADWAY, NEW YORK.
PARIS: J. B. BAILLIERE, LIBRAIRE, RUE HAUTEFEUILLE.
1856.

NOTE.

In this, as in the previous Volumes, the Society is not to be considered responsible for any statements or reasonings offered.

PREFACE.

THE life of Dalton has already been written, but chemical literature seemed to demand a more minute history of the atomic theory up to his time, without at all disparaging the valuable history of chemistry by Dr. Kopp, or the work of Dr. Daubeny, which treats principally of the more modern part. For this and reasons elsewhere mentioned, I have made the distinctive feature of the volume the history of our ideas of matter bearing on modern chemistry, until the time when Dalton flourished. There is a short memoir which breaks off at the fourth chapter, or the time when Dalton first published on atoms, in order to begin the general history, which again leads to Dalton at the eleventh chapter.

Mr. James Woolley was kind enough to lend me all the papers relating to Dalton which he possessed, together with his own MS. memoir, and Mr. Giles, with similar kindness, lent me the memoir which he had written. Dr. Henry's volume also has not been neglected. Mr. Isaac Dickenson, of Cockermouth, and Mr. Thomas Bewley, of Bassenthwaite, have also furnished me with interesting letters.

It might have been expected that more attention would have been given to Dalton's private life, living, as I am, among so many who knew him, but none with whom I have conversed have given any important information not here embodied. I considered also that Dr. Henry had very fully treated that subject, and that it would be unwise and wanting in respect to go over the same ground exactly, even when the same materials were supplied, I have therefore been minute only in such things as did not appear to me elsewhere treated, or such as seemed the most characteristic according to my ideas.

The history of Dalton's many scientific inquiries on subjects other than the atomic theory, is given so shortly that it might almost have been left out, did it not give a greater completeness to the memoir, for the use of such as read no other life of the same man. The history of our ideas of matter is one of the most interesting "fairy tales of science," it is a pity that, like so much else in science and philosophy, it should be so frequently spoiled by dryness. The desire to avoid this has led me to extend further than usual the meaning of the title. The plan of quotation instead of description has been adopted, as the most just as well as the most interesting, if well managed, although one which may gradually be allowed to degenerate into mere *extractive* matter. Every worker and thinker is allowed to speak for himself, and there are few allusions made to the opinions of any, however eminent, who has not himself laboured on the subject. In the court of science every man is his own best witness; by using description instead of quotation we employ advocates who bandy about the meaning, until at last it can nowhere be found. We cannot be too careful of the fame of the absent, were it merely to protest against the loud assertion of self, which is so easy for those who are present, even when their only hope of fame lies in that two-edged truth, that "a living dog is better than a dead lion."

ERRATA.

Near bottom of page 77, the inverted commas should begin, "*For one is not the same.*"

Page 79, for *air and matter*, read *air and water*.

Page 251, for *John Gough W.*, read *Henry Hough W.*

CONTENTS.

MEMOIRS

OF

The Literary and Philosophical Society of Manchester.

MEMOIR OF DR. DALTON,

AND

HISTORY OF THE ATOMIC THEORY.

[*Read October 2nd*, 1855.]

CHAPTER I.

IN 1789, when this Society was first established, chemistry could not with propriety be called a science, although Lavoisier was attempting to decide on some of its more prominent laws, and although Cavendish, Black, and Watt had raised it from that position of obscurity to which the meagreness of its results had so long condemned it, and shewn to the world that it possessed a power, apparently the highest in order. With the exception of these and a very few others, the whole body of its students were under the subjection of one of the strangest delusions that has ever usurped the place of a law of nature. A body of men for many ages at work had made so little progress towards eliciting definite forms of thought upon the elements with which they worked, that the theory of Phlogiston was regarded as a great discovery; a fanciful theory founded on an explanation of facts confessedly incomplete in the eyes of the most enlightened, was made the great centre round which all chemical knowledge placidly rolled. We may be allowed to boast, that, when this society had existed only a few years, one of its members was found to devise a theory which not only was sufficient to throw light on the past, but at once to put into the infinite observations of his predecessors that reason and order of which before they seemed totally deprived. The hermetic mystic and alchemist had toiled over the difficulties, had tried to remove them by physics and by metaphysics, had explained them by supposing an independent will in the elements or by the immediate Divine interposition; by their wildest imaginings and their clearest reasonings little definite had been attained. The laws which Dalton shewed us to be dominant in matter, considered chemically, were at once clear enough to satisfy the most exact reasoner, great enough to satisfy the most poetical thinker, and simple enough to satisfy those who believe that, at least the great primary laws of nature are simple, whether because the highest wisdom can of course attain its ends with the greatest ease, or because the simplest germ is more easily fitted to branch out into an endless development of character and power.

It is scarcely possible to write the life of Dalton without referring to this Society, and as it is by request of this body that I have undertaken to write, I must explain to them what I have specially attempted to do. After Dr. Henry had written the life of Dalton, it might fairly be asked why I should undertake one. I was requested to write this memoir at a time when it was uncertain when Dr. Henry's would appear. I was unwilling to compete with Dr. Henry, and saw no propriety in doing so, although he offered me his materials on certain fair conditions. I had no intention to write a complete life, nor, I believe, had the society the desire to have a memoir so large as that which I present, and I considered it better to confine myself to certain aspects of Dalton's life and discoveries, thus preventing any reason for asking why another memoir should have at all appeared. The portion which I have most carefully worked out is the History of the Atomic Theory. Much has been said of it; some have given the credit to Dalton, some have taken it from him; most writers have confusedly mixed him up with others. Some have looked forward to the probable developments of truth in after times, and undervalued the laws of combination as they now stand, and with them the discoverers. It has been my desire to shew distinctly what of importance each celebrated thinker or worker has said or done in the matter before Dalton, and what he has himself accomplished. This is done by bringing the original words of the authors, and endeavouring to find what amount of meaning can be attached to them. As for all future developments of the present laws I can only say that, believing as I do in the infinite wisdom with which creation is ordered, I am ready to believe in infinite developments of any law; but strange as may be the new and possible combinations of elements, and interesting also the breaking up of these our present elements into other elements, or even into mere pieces, it is not to be imagined that even nature has any thing in store for us fitted to answer the same purpose within the same limits and still simpler and more extensive than the laws of combination as Dalton has expressed them.

Of the man it has been attempted to give only a short sketch, and the whole merely to serve as an enlarged epitaph, written here instead of on his tomb, a token of remembrance for ourselves especially, like the coffin of some departed friend to preside at our feasts, and as a contribution to his defence if he should be assailed.

It will also be more in accordance with his own life if little is said about his personal affairs, which took such a very inferior place in his occupations. Unlike many men of the greatest attainments, Dalton was little occupied with those numerous incidents closely relating to family and friends, which, although productive of much true happiness, capable equally of enlarging the smallest minds and deepening the influence of the most gifted, involve the consumption of much time, a loss much deplored as often as we consider how little we have, and how much is needful in order to obtain from nature even the smallest addition to our knowledge. For Dalton, science was the occupation of life, of a life spent in the most laborious manner. The amenities of life came to him as memories of what had been in his childhood rather than as pleasures realized at the time, memories certainly which he willingly recalled, but as willingly, or perhaps resolutely, left, because his work was before him. He was a student of nature from his cradle.

Few as the materials for his early life are, and bare as are all the narratives, we have perhaps all that could really be found to be interesting. To all appearance he was like those around him, born to be a clodhopper, few things happening to fix the attention of others upon him, no incidents worth a record, because happening to millions daily; the first years must be passed with a simple record of the meagre living and the scantier education he received, until the glow of his life became warmer than that of his fellows. From that time his life is almost entirely in his works, like a devotee who has no heart for the world, but for Divine truth only, his very visits to his early friends were visits as much made for the investigation of truth, made to nature under that aspect which first taught him to observe and to think, which in fact first made him a philosopher and the love of which he never for a moment lost.

DALTON'S EARLY LIFE AND FIRST BOOK.

John Dalton was born at Eaglesfield, in Cumberland, on the 5th of September, 1766, in a small cottage on the estate of the family, which had come into their possession through his maternal grandmother. His father at this time was a woollen weaver, and did not live in the house belonging to the property, but in a cot of his own, having two small rooms below, one some ten feet square only, and the other still less. The larger house was afterwards, on the death of his uncle, occupied by the father. It is described as one of the better class of farm houses of the district, and is still in possession of the family in a somewhat modernized state. The village of Eaglesfield is 2¾ miles south-west of Cockermouth. The whole township contains only 371 inhabitants. His father, Joseph Dalton, was very poor, in fact he earned only a scanty subsistence by weaving common country goods, and his wife eked it out by selling paper, ink, and quills,* but he seems to have been a man of some vigor of mind, as we find that he taught his sons mathematics, giving them such an education at least as is included in mensuration and navigation.† He afterwards inherited the estate, his brother having died childless. He then gave up weaving. It was by a similar occurrence that the property long afterwards came into the possession of Dr. Dalton, and afterwards into the hands of his cousins on the mother's side.‡ Of the Daltons, or the relatives by the father's side, yeomen or small proprietors in Cumberland, we can find little information; his mother, Deborah Greenup, through whom the property came, connected him with many families in the neighbourhood. She was the third daughter of a family of one son and seven daughters, living at Greenrigg, Caldbeck. The son was a barrister, and practised in London, but having inherited Greenrigg, retired to it, living there to an advanced age, having no children, and leaving the property to his unmarried sister, Ruth. This aunt of Dr. Dalton left the estate of

* Mr. Bewley's Letter. † Mr. Woolley.

‡ This estate was of about sixty acres, but Dalton's elder brother Jonathan increased it considerably by purchase.

Greenrigg to be divided equally between Jonathan and John Dalton and their cousin, George Bewley. The estate was small and in a mountainous district, so that when sold in 1817 or 18, it brought only £750.*

I shall give in a note the names of his aunts the Greenups and their families, as their children the cousins of Dr. Dalton were most affectionately looked on by him during his whole life, as well as kindly and liberally remembered in his will.†

Jonathan Dalton, the grandfather, joined the Society of Friends, and the family continued with that body, a circumstance that no doubt in a considerable degree influenced the habits and character of Dr. Dalton.

Joseph Dalton and Deborah Greenup had three children, Jonathan, John, and Mary. The sons were taught mathematics, partly we are told, by the father, but they were also sent to Mr. Fletcher, the teacher of the school belonging to the Society of Friends. Eaglesfield was early connected with this society, and is said to have first built a meeting house for that body. In Mr. Fletcher's school John remained until twelve years of age, and during that time he must have made great progress, as we find him immediately beginning to teach. He always spoke with great admiration of Mr. Fletcher, who lived until Dalton himself was advanced in life. Indeed we have no reason to think that even in that small

* Mr. Bewley's Letter.

† Eldest daughter of Greenup family married to Samuel Bristo; many descendents, particularly Rachel and Margaret Lickbarrow, of Kendal, to whom Dalton left legacies. (There was a third, Isabella, living at Dalton's death.) Second daughter married to Thomas Bewley, whose only son was George Bewley, of Woodhall, who formerly had the school in Kendal. The grandson, Thomas Bewley, of Bassenthwaite, to whose letter I am indebted for a history of the whole family, has daughters married to Mr. Abbatt, of Liverpool, and Mr. Benson, of Preston, who went in the funeral procession as nearest of kin to Dalton, and who by his will received legacies. The fifth married to Mr. Dickinson, whose descendent's letter I have quoted. The sixth married in the north of Scotland; whilst the seventh is mentioned in the text.

place a teacher of ability could not be found. There are other parts of Cumberland, the parish of Martindale for example, where the science of mathematics was well taught in the days of our grandfathers, but where both scholars and instruction in the most elementary English books could with difficulty be had within these three years. Dalton must certainly have surpassed the other scholars when he began to teach at the age of twelve. The school was kept in the Friends' Meeting House, at Eaglesfield, still a school-room.* We are not told if he succeeded his teacher in this, or where Mr. Fletcher afterwards lived. We may picture to ourselves the struggles of the determined boy, working hard at his father's farm in the summer time, as we are informed he did, and helping also to repair the old farm house, but working with still more determination in winter which afforded him the chief opportunities for study, and when the boys from the various farms congregated to the school, their parents not being able to spare them from their work during the busy season of the year. We can picture the indomitable youth, as an old pupil of his, John Robinson, now living at Eaglesfield, has pictured him, struggling for that authority needed to maintain order, but feeling that there was no struggle needed to shew the superiority of his information. Being as old or older than himself they would not be silenced or commanded, and determined as himself they challenged him into the surrounding grave-yard to fight. It is not said whether he accepted the challenge, but he sometimes took the more dignified mode of locking-up the more refractory, repeating in the school-room that they might learn their tasks whilst he went to his dinner. For this, however, he was sometimes at least the greatest sufferer, as they broke the windows in revenge.†

* Pens and ink were announced as to be sold within.

† From Mr. Dickinson's Letter.

Occupied in teaching and in the work of the farm he laid the foundations of an active mind and raised up a vigorous well-knit frame, which underwent great exertion till an advanced age with little interruption from ill health. Here we see the self-reliance which was strong in him through life; at an age when most persons are mere children, he sought to some extent to rule; and when most persons have scarcely begun to learn soundly, he sought to teach. Here, also, we see that peculiarity of his mind, which did not seek to acquire a great mass of information otherwise than by investigation, and had more pleasure in making use of what it had attained either by conveying it to others, or as a tool for search. These united causes throw some light on his early grasp at independence, as it was not necessity that compelled him to work, nor the want of the means of living which had never failed him.

As we are obliged to arrive at his early character chiefly by inference, we must the more carefully remember, what is more directly told of him, his great diligence. This continued with him through life, and his theory of success was made in the belief that diligence constituted the main difference amongst men engaged in intellectual pursuits. His principal study in early life was mathematics, which he learned with a boy of the name of William Balderstone, receiving assistance from a gentleman in the neighbourhood of the name of Robinson, who fortunately had education as well as property. He and his accomplished wife readily gave their assistance in conducting the studies of Dalton, and his companion who was in their service. These boys filled with emulation solved problems in numbers and in forms, as active minded boys still do over all the country. Balderstone bet Dalton sixpence on a proposition in geometry, but Mr. Robinson objected, and proposed rather that whoever lost should supply the other with candles during the winter. Dalton's answer is generally said to be, "yan might do it," for although his parents had got over one of their early difficulties, that of "keeping him out of the dubs," they had not yet taught him to speak other than his native dialect, which he not unfrequently used also in later life, thereby giving pith or humour to his conversation. We scarcely know if the mention of "dubs" means that he was fonder than other children of such things, or whether his love of nature first took its rise in this common although unpromising amusement.

This acquaintance with Mr. Robinson began when Dalton was about ten years old, about the age when he solved a problem, discussed on a hay field among the farm people, whether sixty yards square or sixty square yards are the same. He at first considered them the same, but reflection shewed him the difference. Dalton seldom failed for want of perseverance; he cheered his weary companion who was soon outstripped, and who lived to look on the youth to whom he supplied candles, with the reverence of those who deeply conscious of ignorance imagine in knowledge the most extravagant powers.

Dalton insisted on the importance of diligence, without however considering that the work on which his fame was founded was done comparatively in early life, and that his subsequent unwearied application in no ways tended to elevate his position in science. There is seldom much fame for the idler, but we err greatly when we say a word to dishonour the greatest of all gifts, which cannot be called by a less name than Divine, the eye of genius.

On leaving the boyhood of Dalton we are not called to look on it with surprise, we see in it indications of force, but an equal display is sometimes apparent in less gifted men. We can scarcely look with wonder at the elevation we have seen him attain above his humble fellow-students and pupils. Self-cultivation, too, is a problem now happily so often solved by those who have nothing of true genius, that we can only look on his acquirements at that time as àttainable with the greatest ease by one who had evidently that gift in great force.

After spending about three years teaching in this school he left Eaglesfield. About this time, getting on in life as he thought, he first saw an umbrella in Cockermouth, and bought one, thinking, as he afterwards expressed it, that he was now becoming a gentleman. This happened in 1781, when he became assistant to his cousin, George Bewley, who kept a school in Kendal. His brother Jonathan had been there as assistant for some time previously. In this place twelve years of his life were spent, and here his true education began. He had learned some Latin and Greek, but neither now nor at any time of his life does he seem to have attended much to literature or philology. He is said to have had such an excellent memory that he repeated some of Anacreon's Odes forty years after he had read them, but a few of Anacreon's musical lines seem to be the common property of school boys, who learn them easily from their sound, whilst this knowledge gives no indication whatever of proficiency in the language. A few old Greek books were sold with the rest of his library, partly his own and partly his brother's small stock, at Kendal. The Greek dictionary, a Schrevelius, seems never to have been used. Reading Greek books was no sport for a man who made forty thousand meteorological observations.

In Kendal he became acquainted with Mr. Gough, a man who although blind from infancy, was possessed of high scientific attainments. The mutual assistance rendered is best expressed in Dalton's own words. "For about eight years during my residence in Kendal, we were intimately acquainted; Mr. Gough was as much gratified with imparting his stores of science, as I was in receiving them: my use to him was chiefly in reading, writing, and making calculations and diagrams; and in participating with him in the pleasure resulting from successful investigations: but as Mr. Gough was above receiving any pecuniary recompense, the balance of advantage was greatly in my favour; and I am glad to have this opportunity of acknowledging it. It was he who first set the example of keeping a meteorological journal at Kendal." Meteor. Observ. and Essays. Preface, 1834.

In an earlier edition Dalton had mentioned Gough as an unknown friend; in the quotation given he does him that justice which in his lifetime he was forbidden to do. Mr. T. T. Wilkinson mentions that Dr. Whewell and several other distinguished wranglers were prepared for their contests by Mr. Gough. He himself wrote no separate work, but many of his scientific memoirs have appeared. As a man he seems to have inspired great respect in all who knew him, and he must have been no common person of whom Wordsworth wrote;

Methinks I see him how his eyeballs roll'd
Beneath his ample brow, in darkness pained
But each instinct with spirit, and the frame
Of the whole countenance alive with thought,
Fancy, and understanding; whilst the voice
Discoursed of natural or moral truth,
With eloquence and such authentic power,
That in his presence humbler knowledge stood
Abashed, and tender pity overawed.

The Excursion.

The poet, in a letter to Mr. Samuel Crompton, of Manchester, writes, "Your conjecture concerning that passage is remarkable; Mr. Gough, of Kendal, whom I had the pleasure of knowing, was the person from whom I drew the picture, which was in no respect exaggerated. He was a most extraordinary person, highly gifted, &c. The sadness which the contemplation of blindness always produces, was in Mr. Gough's case tempered by admiration and wonder in the most affecting manner."

During the time that Dalton was enjoying the instruction and advice, as well as the library and scientific apparatus of

Mr. Gough, his cousin George Bewley gave up the school, and the two brothers who had assisted him thus announced their intention of continuing it.

"Jonathan & John Dalton respectfully inform their Friends, and the Public in General, that they intend to continue the school taught by George Bewley, where Youth will be carefully instructed in English, Latin, Greek, & French; also Writing, Arithmetic, Merchant's Accounts, and the Mathematics. The school to be opened on the 28th of March, 1785.

N.B. Youth boarded in the Masters' house on reasonable terms."

Their sister Mary came to keep the house, and their father and mother, then old people, frequently went to see them, walking through on one day over "mountain and slack" a distance of 45 miles.*

George Bewley lent them some money, probably the house or furniture answering to the sum, which was returned the same year; the father lent them seven guineas, "to be paid 9 mo. 29th," which they paid only a week behind time. And so they began life on a larger scale. They took care of their money, balanced their books every month, and put down every penny they spent. They had more than once to get a guinea from Mr. Lickbarrow and two from Mr. Kendal, and Mary had to give up her thirteen shillings and sixpence, and got paid in portions, "Mary, in part, 0. 0. 6d." Mr. Benson too was paid all his money, and borrowings soon ceased. The whole sum got the first year was about £107, but a good deal of this had to be paid back, and indeed the average of the school was about seventy pounds a year. This was increased a few pounds by "drawing conditions," "collecting rents," "making wills," and "searching registers," but the amount gained by this means was seldom above five pounds a

* John Robinson, in Mr. Dickinson's Letter.

year, although it rises to more than twice that sum in certain years. We do not know if Dalton so employed himself, certainly however his brother did so.

A second circular issued by the Daltons in the following year shews more fully the intentions of the school, and gives us also an idea of the enlarged views of education which Dalton took.

Kendal, July 5th, 1786.

"Jonathan & John Dalton take this method of returning their acknowledgments to their friends and the public for the encouragement they have received since their opening school; and from their care and assiduity in the management of it manifested in the improvement of the Youth under their care, are induced to hope for a continuation of their favours. They continue to teach on reasonable terms English, Latin, Greek, and French;

Also,

Writing,	Mensuration,	Projections,
Arithmetic,	Surveying,	Dialling,
Merchant's Accounts,	Gauging,	Optics,
Geometry,	Algebra,	Mechanics,
Trigonometry,	Fluxions,	Pneumatics,
Navigation,	Conic Sections,	Hydrostatics,
Geography,	Astronomy,	Hydraulics, &c.

N.B. Persons desirous of being instructed in the use of the globes, &c., will be waited upon any time out of school hours. The Public may also be informed, that they could conveniently teach a considerable number more than at present. Those who send their children may depend upon their being carefully instructed."

To his usual instruction in school he added lectures. It is certainly interesting to look over the syllabus published. We see him as an ardent young man, filled with ideas from every science, eager to tell them to others. We must remember that he is a man of simple mind, who had never seen a large city, to whom the parade of the world and the unfortunate pride of successful science was known only from books, and we need not wonder that he expected on announcing the truths he had learned to an ignorant rural population, a crowd eager to hear them. Independent investigation had not yet contracted his field of interest, so to speak, for we find this effect produced of necessity in the mind which has for a long time travelled alone over an unmapped district. He is now 21 years of age, and gives the following programme of lectures.

Oct. 26th, 1787.

"Twelve lectures on Natural Philosophy to be read at the school (if a sufficient number of subscribers are procured) by John Dalton. To begin on Tuesday evening the 13th Nov$^{r.}$ next, at 6 o'cl., and to continue every Tuesday and Thursday at the same hour till compleated.

Subscribers to the whole ½ a guinea; or one shilling for single nights.

N.B. Subscribers to the whole course will have the liberty of requiring further explanation of subjects that may not be sufficiently discussed or clearly perceived when under immediate consideration; also of proposing doubts, objections, etc.; all which will be illustrated and obviated at suitable times to be mentioned at the commencement.

A Syllabus of the Lectures.

First & Second. Mechanics.

Introduction. Rules of Philosophizing on Matter and its Properties with the different opinions of the most famous Philosophers on this head.

The laws of motion. Mechanic powers. Vibration of pendulums.

Third, Fourth, & Fifth. Optics.

Preliminary discourse. Of the nature and properties of light. Of simple vision. Doctrine of colours. Of reflected vision. Of mirrors and images reflected from them. Of refracted vision, with the nature of lenses and images exhibited thereby. Of burning glasses. Description of the eye. Manner of vision. Of long and short sighted eyes. Of spectacles, telescopes, & microscopes. Of the rainbow.

Sixth & Seventh. Pneumatics.

Of the atmosphere. The elasticity of the air. Description of the air pump. The spring and weight of the air proved by a great variety of experiments on the air pump. Of respiration. Of sound. Of winds. Of the blueness of the sky. Of twilight.

Eighth, Ninth, & Tenth. Astronomy.

Introduction. Of the solar system. Of the figures, magnitudes, distances, motion, &c. of the sun, planets, and comets. Of the progressive motion of light. Of the fixed stars and their phenomena. Of the lunar planets. Of eclipses, tides, &c.

Eleventh & Twelfth. Use of the Globes.

Figure of the Earth. Description of the globes. Various problems performed thereon, amongst which are, an explanation of the phenomena of the harvest-moon and the variations of the seasons. Conclusion.

'Ex rerum causis supremam noscere causam.'"

Miss Johns, whose diary will be spoken of below, tells us that this very syllabus and one for 1792 came accidentally in his way in after-life when he was looking over some old letters, having been detained in the house by a cold. He burst out into a loud laugh.

The accounts given of the Daltons as teachers lead us to believe them to have kept up the old system of great sternness and formality, although John's character seems to have been the milder of the two. Even during school hours he was much occupied with mathematics and making calculations at all spare moments on bits of paper that came in his way. This sternness of manner never left him, although his disposition was undoubtedly gentle. It may have arisen from having been compelled so long to continue his pursuits without the companionship of congenial minds, although even that does not sufficiently explain the cause, as we hear of others less befriended than Dalton whose disposition has retained its vivacity. It is greatly to be regretted that a journal containing all the minute particulars of his life at this period should be entirely lost. Portions which have been read to me by his friend Peter Clare introduce us into his character in a very pleasing manner. We find him cheerful and easy, fond of a little innocent sport, and much attached to some games, but still so precise that every one was rigidly recorded and the results of the play of each party systematically compared. One evening at a house he visited, the company spent their time making verses; when the last word of one verse was told, the next person in order was expected to make a line to rhyme to it. It is curious to observe that every couplet, as well as the author of each, is carefully noted down in the diary. In this we have an early illustration of the great order that was a prominent point to be remembered in judging of his intellectual character. This does not deny the paradoxical addition of great carelessness. It was at this time that he was more especially a student of the Lady's Diary, and one of those who solved its problems, obtaining on several occasions the prize. Even at Eaglesfield, however, he was employed, although not with equal success, in the same manner, as Mr. Dickinson says, "When I was a boy I saw John Dalton at my cousin William Alderson's house, in Eaglesfield; they talked of days when they were lads together, sitting over the fire till midnight poring over the Lady's Diary. John never giving sleep to his eyelids until he had found out the riddle of some prize enigma or some mathematical question."

We find him at this time making his own barometers and

thermometers, making experiments on hygrometers of whip cord, sending specimens of butterflies to Mr. Crosthwaite's museum in Keswick, and afterwards engaging for a very small sum to send him dried specimens of plants. Mr. T. P. Heywood, of the Isle of Man, has in his possession eleven volumes of his *Hortus Siccus.* The first is a thick volume, containing the general title page, *Hortus Siccus seu Plantaram diversarum in Agris Kendal vicinis sponte nascentium Specimina, Opere et Studio Johannis Dalton collecta, et Secundum classes et ordines disposita.* 1790." The other volumes are thin. They are not preserved with the greatest care which collectors are capable of, but they are a proof of great industry and of considerable attainment, even in a branch of science which he did not profess. He supplied Mr. Crosthwaite with a barometer and thermometer, although not knowing how to make them. He nevertheless begins and learns their faults by experience. A letter given by Dr. Henry shews us how his knowledge stood at this time, and how also he was in the habit of acquiring it. Speaking of the mercury in the barometer, he says, "I intend to renew mine as soon as convenient; if thou do the same, be careful in undoing it; and attend to the cautions I give. Be sure to rub the inside of the tube well with warm dry cotton or wool, and have the mercury when poured in at least milk warm, for moisture is above all things else to be avoided, as it depresses the mercury far more than a particle of air does; mine is, as I have said, at least $\frac{1}{10}$th of an inch too low, and yet it is clear of air, and to all appearance dry; but I doubt not but attending to these precautions, which I knew nothing of when it was filled, will raise it up to its proper height."

At this time he felt uneasy, the sphere of his simple school was too small, his impatience took the form of variety in his pursuits, and he wandered over nearly every branch of science. He seems now, although at a later time than generally happens to young men, to have been "yearning for that large" existence which he knew to be somehow attainable. He thought of a profession. His uncle Thomas Greenup, the barrister, thought that it was entirely out of the reach of a person in Dalton's circumstances to be a barrister or a physician, and recommended rather as less difficult, but still much above him, that of an attorney or apothecary. It was at this time that he began to make medical experiments, wishing to ascertain the loss from the human body by insensible perspiration, a sufficient proof that however much he might have advanced the knowledge of that profession, he was too much an experimenter and solitary thinker to have been pleased with the active life of some of our medical men. The discouragement received from his friends seems to have prevented all exertion in the new direction he had contemplated, and he remained three years longer or until 1793 in Kendal, when Dr. Barnes asked Mr. Gough for a suitable person to teach mathematics in the New College of Manchester. This college had arisen out of the Warrington academy, where Priestley had taught, as well as Dr. Aikin, Dr. Enfield, Reinhold Forster, and Gilbert Wakefield. At that time the college was in the present "College-buildings," in Mosley-street. He lived in the establishment, and remained tutor of mathematics and natural philosophy for six years. Dr. Barnes was the principal. This college is now transferred to Gordon-square, London. Whilst here we find from papers lent me by Mr. Woolley, that in 1794 he had twenty-four pupils for mathematics, mechanics, geometry, algebra, book-keeping, natural philosophy, and chemistry. He used Lavoisier's Elements of Chemistry and Chaptal's amongst others. In 1799 there were twenty-two students. Although Manchester is now multiplied by four, it cannot shew the same number, and I fear that the love of external things has overpowered the love of science.

As soon as he gave up the intention of studying for a profession, he seems to have decided at once on a regular course of investigation and in the interval from that time to 1793, when he went to Manchester, produced his first work properly so called, his "Meteorological Observations and Essays;" they were not, however, published until he had taken up his position at the college.

This book contains an extensive series of observations on old and new subjects, comprising ideas sometimes new, sometimes old, and at other times modifications of the old. He enters into the discussion of the cause of the rise and fall of the barometer, which he decides to be the existence of the vapour of water in the air. Also he discusses the state of water in the atmosphere, shewing it to be an elastic vapour existing like any other gas not in chemical combination. Then he treats of evaporation from the earth's surface, clouds and rain, and allied phenomena, bringing, as Professor Sedgwick says, "the elements themselves under his own intellectual domination." The extent and variety of these inquiries prove the earnestness with which he studied in his almost solitude in Cumberland. The work seems to have been at first intended as a popular treatise on meteorology. It begins with a description of the barometer, then come the thermometer, hygrometer, and rain gauges; connected with these are tables of observations made at Kendal and Keswick. There seems to be a looseness of description in the first part of the volume, which would seem to imply that the matter was easily understood, and the readers could make out the particulars for themselves. As he proceeds, however, he seems to feel that he has a harder task to perform, and speaks rather to scientific than to popular hearers, whilst we gradually become aware that he is a close and precise reasoner. His style is very simple; he goes directly to his point; all inessential parts are left out. He seems to move forward with a heavy dogged tread, never turning his head aside, but as any style may become a fault if too far carried out, we find that in his there are left out many things that are certainly needful as accessory or confirmatory, leaving what to the eyes of many is a want of finish, so that others have been needed to complete what was in reality sufficiently complete had it been laid out entire as it existed in his own mind.

As an example of his style, at p. 97, we find

"It appears from the observations (see table p. 15) that the mean state of the barometer is rather lower than higher in winter than in summer, though a stratum of air on the earth's surface always weighs more in the former season than in the latter; from which facts we must unavoidably infer that the height of the atmosphere, or at least of the gross parts of it, is less in winter than in summer, conformable to the table p. 80. There are more reasons than one to conclude that the annual variation in the height of the atmosphere, over the temperate and frigid zones, is gradual, and depends in a great measure on the mean temperature at the earth's surface below, for clouds are never observed to be above four or five miles high, on which account the clear air above can receive little or no heat, but from the subjacent regions of the atmosphere, which we know are influenced by the mean temperature at the earth's surface; also, in this respect, the change of temperature in the upper parts of the atmosphere must in some degree be conformable to that of the earth below, which we find by experience increases and decreases gradually each year, at any moderate depth, according to the temperature of the seasons.

"Now with respect to the fluctuations of the barometer, which are sometimes very great in twenty-four hours, and often from one extreme to the other in a week or ten days, it must be concluded, either that the height of the atmosphere over any country varies according to the barometer, or otherwise that the height is little affected therewith, and that the whole or greatest part of the variation is occasioned by a change in the density of the lower regions of the air. It is very improbable that the height of the atmosphere should be subject to such fluctuations, or that it should be regulated in any other manner than by the weekly or monthly mean temperature of the lower regions; because the mean weight of the air is so nearly the same in all the seasons of the year, which could not be if the atmosphere was as high and dense above the summits of the mountains in winter, as it is in summer. However, the decision of this question need not rest on probability; there are facts which sufficiently prove, that the fluctuation of density in the lower regions has the chief effect on the barometer, and that the higher regions are not subject to proportionable mutations in density. In the Memoirs of the Royal Academy at Paris, for 1709, there is a comparison of observations upon the barometer, at different places, and amongst others, at Zurich, in Switzerland, in lat. 47° N., and at Marseilles, in France, lat. 43° 15′ N.; the former place is more than 400 yards above the level of the sea; it was found that the annual range of the barometer was the same at each place, viz., about 10 lines; whilst at Genoa, in latitude 44° 25′ N., the annual range was 12 lines, or 1 inch; and at Paris, latitude 48° 50′ N., it was about 1 inch 4 lines. In the same memoir it is related that *F. Laval* made observations, for ten days together, upon the top of St. Pilon, a mountain near Marseilles, which was 960 yards high, and found that when the barometer varied $2\frac{3}{4}$ lines at Marseilles, it varied but $1\frac{3}{4}$ upon St. Pilon. Now had it been a law, that the whole atmosphere rises and falls with the barometer, the fluctuations in any elevated barometer would be to those of another barometer below it, nearly as the absolute heights of the mercurial columns in each, which in these instances were far from being so. Hence then it may be inferred, that the fluctuations of the barometer are occasioned chiefly by a variation in the density of the lower regions of the air, and not by an alternate elevation and depression of the whole superincumbent atmosphere. How we conceive this fluctuation in the density of the air to be effected, and in what manner the preceding general facts relative to the variation of the barometer may be accounted for, is what we shall now attempt to explain." This is referred to the varying amount of vapour.

In section 5, "observations on the height of clouds," there is given the summary of 5381 observations, made by Mr. Crosthwaite, an evidence of the intellectual diligence to be found at the lakes even before they became the haunt of poets.

At p. 127 he gives a table of the temperature of water made to boil at different atmospheric pressures, bearing on the fact, which he is there explaining, that "aqueous vapour always exists as a fluid sui generis diffused amongst the rest of the aerial fluids;" and at p. 129, "that it may be determined a priori what weight of vapour a given bulk of dry air will admit of, for any temperature, provided the spec. grav. of the vapour be given." These conclusions appear more fully in a note* to a paper read in 1797, after having made confirmatory experiments. This must be taken as an elucidation of a subject which afforded much discussion at that period. We know that Saussure and others knew well that moisture existed in the air at very low temperatures, and there was a variety of opinions as to the state in which it existed. Many writers of the period believed, that because warm air was sensibly drier, it contained less moisture than cold air; all these points Dalton has elucidated and spoken on with decision.

In the third and sixth essays, the precise point where the burden of his argument is contained is difficult to find, the reasoning being a constant process to prove what it is supposed we knew beforehand to be the result. In the appendix, he says more pithily, p. 188, "I am confirmed in the opinion that *the vapour of water, and probably of most other liquids, exists at all times in the atmosphere, and is capable of bearing any known degree of cold without a total condensation, and that*

* Memoirs of the Manchester Philosophical Society. Vol. V., p. 351.

the vapour so existing is one and the same thing with steam, a vapour of the temperature of 212° *or upwards.* The idea, therefore, that vapour cannot exist in the open atmosphere under the temperature of 212°, unless chemically combined therewith, I consider as erroneous; it has taken its rise from a supposition, that air pressing upon vapour condenses the vapour equally with vapour pressing upon vapour, a supposition we have no right to assume, and which I apprehend will plainly appear to be contradictory to reason and unwarranted by facts; for, when a particle of vapour exists between two particles of air, let their equal and opposite pressures upon it be what they may, they cannot bring it nearer to another particle of vapour, without which no condensation of vapour can take place, all other circumstances being the same; and it has never been proved that the vapour in a receiver from which the air has been exhausted, is precipitated upon the admission of perfectly dry air. Hence, then, we ought to conclude, till the contrary can be proved, *that the condensation of vapour exposed to the common air does not in any manner depend upon the pressure of the air.*"

At p. 135, after contending for the theory that the vapour of water is mixed with the air and not combined, he explains how it is precipitated by cold and taken up by heat, and how it is that clouds consisting of light drops do not fall so readily as clouds with heavy drops, as the resistance of the drops is as the square of the diameter, in which his mathematical knowledge helps his meteorology. This was suggested by Mr. Gough.

There is in these Essays, and everywhere in Dalton's writings, a great rapidity of reasoning, a direct passage from premise to conclusion without fear, as if more than usually persuaded that true reason could not misguide him, so that he is utterly regardless of consequences.

At p. 168 we find a fair example of his mode of reasoning, and one also of his daring theories.

"The light of the aurora has been accounted for on three or more different suppositions:—1. It has been supposed to be a flame arising from a chymical effervescence of combustible exhalations from the earth. 2. It has been supposed to be inflammable air, fired by electricity. 3. It has been supposed electric light itself."

"The first of these suppositions I pass by as utterly inadequate to account for the phenomena. The second is pressed with a great difficulty how to account for the existence of *strata* of inflammable air in the atmosphere, since it appears that the different elastic fluids intimately mix with each other; and even admitting the existence of these *strata*, it seems unnecessary to introduce them in the case, because we know that discharges of the electric fluid in the atmosphere do exhibit light, from the phenomenon of lightning. From these and other reasons, some of which shall be mentioned hereafter, I consider it almost beyond doubt that the *light* of the *aurora borealis*, as well as that of falling stars and the larger meteors, is electric light solely, and that there is nothing of combustion in any of these phenomena. Air, and all elastic fluids, are reckoned amongst the non-conductors of electricity. There seems, however, a difference amongst them in this respect: dry air is known to conduct worse than moist air, or air saturated with vapour. Thunder usually takes place in summer, and at such times as the air is highly charged with vapour; when it happens in winter, the barometer is low, and consequently, according to our theory of the variation of the barometer, there is then much vapourized air; from all which it seems probable, that air highly vapourized becomes an imperfect conductor, and, of course, a discharge made along a stratum of it will exhibit light, which I suppose to be the general cause of thunder and lightning."

"Now, from the conclusions in the preceding sections, we are under the necessity of considering the *beams* of the *aurora borealis* of a *ferruginous* nature, because nothing else is known to be magnetic, and consequently, that there exists in the higher regions of the atmosphere an elastic fluid partaking of the properties of *iron*, or rather of *magnetic steel*, and that this fluid, doubtless from its magnetic property, assumes the form of cylindric beams. It should seem, too, that the rainbow-like arches are a sort of *rings* of the same fluid, which encompass the earth's northern magnetic pole, like as the parallels of latitude do the other poles; and that the beams are arrayed in equidistant rows round the same pole. * * * Things being thus stated, I moreover suppose that this elastic fluid of magnetic matter is, like vapourized air, an imperfect conductor of electricity; and that when the equilibrium of electricity in the higher regions of the atmosphere is disturbed, I conceive that it takes these beams and rings as conductors, and runs along from one quarter of the heavens to another, exhibiting all the phenomena of the *aurora borealis*."*

In the edition of 1834 he still adheres to the same theory; some will look on it as absurd; it is certainly the result of great daring, or in other words, it may be viewed as the reasoning of a man who has exhausted all his knowledge in finding a cause, feels certain that there is one, and decides upon that which is most conformable to his knowledge, without waiting for a wider view, or for a time when something perfectly new might entirely change the scene.

This essay on the aurora he considered as of great importance. He begins with these words; "As this essay contains an original discovery which seems to open a new field of inquiry in philosophy, or rather, perhaps, to extend the bounds of one that has been as yet but just opened; it may not, perhaps, be unacceptable to many readers to state briefly the train of circumstances which led the author to the important conclusions contained in the following pages." And yet we take up treatises on the aurora, and do not even find Dalton's name

* The pages refer to the new edition.

mentioned. Wargentin, Halley, and Celsius had all observed the action of the aurora on the magnetic needle. Dr. Halley had supposed it to be caused by magnetism. Dalton went more fully into the subject than his predecessors, without, however, taking all difficulty from it. In this treatise we see an instance of the pertinacity with which he held ideas which he had formed. But we find him altering his opinion on the height of the aurora; his observations led him to believe the height to be about 150 miles; afterwards he considered it to be about 100. Numerous as have been the attempts to ascertain the height, the differences ranging from feet to thousands of miles, Dalton still, in 1834, severely criticised all the observations which differed greatly from his early results. To this treatise on meteorology he added little, although a new edition appeared after forty years. He then says, that it is printed verbatim (adding only a small appendix), "as I apprehend it contains the germ of most of the ideas which I have since expanded more at large in different essays, and which have been considered as discoveries of some importance." But he says also, that "the subject here treated of appeared to the author to be very imperfectly appreciated, or little understood, by some of the modern writers on meteorology," and it is probably true that the facts and theories he advanced had, in some or many instances, been worked out by others with little aid from his book, because, although occasionally quoted, it was really very little known. This arose from a peculiarity in his mode of publishing it. It was like all his books printed for himself, and was never allowed to make its due way in an independent manner among the booksellers, nor had the essays the advantage of being read to a society, or given out by any journal.

CHAPTER II.

HIS EPOCH OF GREATEST FERTILITY.

THE work on meteorology was published in September, 1793. Dalton had come to Manchester in the spring of that year. He was now twenty-seven years of age. The removal to an active town seems to have satisfied his cravings for a larger sphere of labour which were forcing him from his attachment to his neighbourhood. He was self-taught, a raw countryman, in many respects rather rough in his acquired habits, although of a naturally gentle disposition. Such a distance from active life would have made many men idle, such a sudden entrance into it has often the same effect on others. Neither seemed to affect him, there was little change of habit, he was still in the streets of Manchester as on the hills of Cumberland, the active observer and thinker. On October 3rd, 1794, he first appears as a member of the Literary and Philosophical Society of Manchester, having been proposed by Thomas Henry, Dr. Percival, and Robert Owen, the veteran enthusiast who would willingly compel all mankind to be reformed by his simple formula. On the 31st, he read his first paper to the society, an event to him of great importance, greatly influencing all his future life, as he soon after became the representative of that body, continuing so for the remainder of his life.

This paper was entitled "Extraordinary Facts relating to the Vision of Colours."* He says there, p. 30,

"It may be proper to observe, that I am shortsighted. Concave glasses of about five inches focus suit me best. I can see distinctly at a proper distance; and am seldom hurt by too much or too little light; nor yet with long application."

* Memoirs of the Philosophical Society of Manchester. Vol. V., p. 28.

"I found that persons in general distinguish six kinds of colour in the solar image; namely, red, orange, yellow, green, blue, and purple. To me it is quite otherwise; I see only two, or at most three, distinctions; these I should call yellow and blue, or yellow, blue, and purple. * * * * My yellow comprehends the red, orange, yellow, and green of others, and my blue and purple coincide with theirs."

He sums up the peculiarities of the vision of himself and others who have been found similarily affected thus; p. 40.

"1. In the solar spectrum three colours appear—yellow, blue, and purple. The two former make a contrast; the two latter seem to differ more in degree than in kind.

2. *Pink* appears, by day light, to be sky-blue a little faded; by candle light it assumes an orange or yellowish appearance, which forms a strong contrast to blue.

3. *Crimson* appears a muddy blue by day; and crimson woollen yarn is much the same as dark blue.

4. *Red and scarlet* have a more vivid and flaming appearance by candle light than by day light.

5. There is not much difference in colour between a stick of red sealing wax and grass, by day.

6. Dark green woollen cloth seems a muddy red, much darker than grass, and of a very different colour.

7. The colour of a florid complexion is dusky blue.

8. Coats, gowns, &c., appear to us frequently to be badly matched with linings, when others say they are not. On the other hand, we should match crimsons with claret or mud; pinks with light blues; browns with reds; and drabs with greens.

9. In all points where we differ from other persons, the difference is much less by candle light than by day light."

He found several persons having the same peculiarity of vision, and says (p. 43), "It appears, therefore, almost beyond a doubt, that one of the humours of my eye, and of the eyes of my fellows, is a coloured medium, probably some modification of blue."*

Although this paper was an observation on himself, it is in reality a discovery; the facts had not been arranged before he arranged them, and found out other persons similarly situated. A peculiar keenness of reasoning was needed to find it out, as we must remember that with such persons there is little community of feeling on colour, and scarcely a mode of judging whether they see any colours exactly as the normal eye does. It would probably explain many strange occurrences if we were to consider that there are really persons in the world who see all crimsons as "dark blue" or "a muddy blue," and who would "match crimsons with claret or mud; pinks with light blues; browns with reds; and drabs with greens;" who see the healthful tints of a florid complexion to be like "dilute black ink on white paper," or "a dull opake blackish blue, upon a white ground." How many strange mistakes and visions might be accounted for by this defect of sight. A fair face with glowing veins would be to Dalton as a corrupting corpse. But it may be said that custom would make all appear as well to him as to others; no, it cannot be so: a defect must constantly carry with it the consequences of a defect, and in this case the established difference which nature has made between life and death, beauty and horror, was hidden from the eye, and therefore to a great extent must have been concealed from the intellect. To this cause partly we may refer that want of fine sensibility to external things which peculiarly marked his scientific as well as social life.

Dr. Whewell has called such persons *idiopts*, because their vision is peculiar; this is not sufficiently characteristic, and

* Mr. J. A. Ransome, who examined the eye after death, found nothing whatever to account for the peculiarity of vision. Certainly colours appeared as usual through it. He believed that the cause was a deficient sensorial or receptive power.

as has been remarked sounds badly, Sir John Herschel having changed it to Dichromic vision, believing that one of the three colours is lost to the eye entirely. Such a vision there seems to be, but this extent has not been observed in any instances, by Dr. George Wilson, who thinks that there is no colour quite lost, although the power of perceiving be feeble, and he names it *Chromato-Pseudopsis*, or a false vision of colours. This he has translated by Sir D. Brewster's term, colour-blindness, which appears much too strong when we consider that some colours are well seen, and others seen in part. It seems, in fact, to be an imperfection in the power of distinguishing colours, which may exist to any extent, either very slightly, as is seen in every-day life, where, for example, among the many workpeople in a large mill, only a few are found fit for arranging yarns with accuracy. A nice perception of colour is there a valuable gift, and is paid for accordingly. Or it may occur decidedly defective, as with Dr. Dalton and others. Dalton's brother had the same defect, and one or two others in the neighbourhood of Eaglesfield, of whom I have lately heard. It is probable that there are many gradations, beginning with deficient colour sight and ending in Dichromic, or perhaps Monochromic or Achromic vision, or true colour blindness. Dr. Wilson well remarks, that Daltonism, under which it has been known, is not a proper name for the peculiarity, as it connects his name with a defect. Indeed few eyes are found equal to Dalton's, if we judge of them by their results. Dr. Wilson has made the remarkable discovery that this defect may almost be called common.

Dalton remained without giving anything to the public until 1799. In the College his order showed itself in the careful list of students and their lessons, still remaining. Possibly his duties occupied too much of his time to allow of experiment, but he comes out so suddenly after that as physikist and chemist, that his time must have been spent in suitable studies. On March 1st, 1799, he read to the Literary and Philosophical Society* "Experiments and observations to determine whether the quantity of rain and dew is equal to the quantity of water carried off by the rivers and raised by evaporation; with an inquiry into the origin of springs." In this he treats,—

"1. Of the quantity of rain and dew.
2. Of the quantity of water that flows into the sea.
3. Of the quantity of water raised by evaporation.
4. Of the origin of springs."

The first three are accompanied by experiments, but there is a looseness in the calculations which renders the paper rather like a sketch of the subject. He, however, collects a great deal of information as to the annual fall of rain in various places, and in a note explains clearly, as before alluded to, his ideas as to the state of aqueous vapour in the air. The looseness of expression is not at all times with him an indication of want of decision, but his peculiar style of writing, as if every one knew the subject, and were ready to draw out his reasoning into all its details, as soon as expressed. His experiments, begun with the hand, seem often finished with the head, so rapidly are his conclusions come to, and the natural law established in his mind. Even now we can add little to the relation between evaporation, rain, and dew, and on the origin of springs he is clear, quick, and decisive, saying that they come from the rain. This subject had been much disputed; filtration from the sea having been a favourite method of obtaining the water, as well as subterranean reservoirs like those of Father Kircher, who shows them in engravings continually boiling out from the centre of the earth. Dalton was not the first to suggest the explanation, of course, but the subject was sufficiently uncertain to call for elucidation. On April 12th,

* Memoirs of the Literary and Philosophical Society of Manchester. Vol. V., p. 346.

1799, he read a paper entitled "Experiments and observations on the power of fluids to conduct heat, with reference to Count Rumford's seventh essay on the same subject."*

He seems evidently to have made up his mind at once that Count Rumford had drawn a wrong conclusion from his premises, and we see in the reasoning much minute ingenuity and acuteness. As an example of his mode of experimenting and reasoning, the following may be given, p. 381.

"Exp. 3. Took an ale glass of a conical figure, 2½ inches in diameter, and 3 inches deep; filled it with water that had been standing in the room, and consequently of the temperature of the air nearly. Put the bulb of a thermometer to the bottom of the glass, the scale being out of the water; then having marked the temperature, I put the red-hot tip of a poker half-an-inch deep into the water, holding it there steadily about half-a-minute; and as soon as it was withdrawn, I dipped the bulb of a sensible thermometer about ¼ inch, when it rose in a few seconds to 180.°

TEMPERATURE.

Time.	At Top.	Middle.	Bottom.
Before the poker was immersed	...	...	47°
—	180°	—	47°
5 min.	100°	60°	47½°
20 ,,	70°	60°	49°
1 hour	55°	—	52°"

After other experiments he says, p. 385, "We must conclude, therefore, that the quick circulation of heat in water over a fire, &c., is owing *principally* to the internal motion excited by an alteration of specific gravity; but not *solely* to that cause, as Count Rumford has inferred."

A very simple and ingenious experiment is related on the same page. He mixed hot and cold water, stirred for half-a-minute, and tried if the upper part became hotter than the lower, it

* Same vol., p. 373.

did not do so, on which he says, "If the particles of water during the agitation had not actually communicated their heat, the hot ones ought to have risen to the top, and the cold ones subsided, so as to have made a material difference in the temperature." This shows, that even at that period he was accustomed to think habitually of matter as decidedly atomic in its constitution.

On the theoretical conclusion to be drawn here, we find his genius taking the lead; he is accurate in spite of the rudeness of his experiments. He concludes that water conducts heat a little, and that the expansion of water is the same both above and below the point of maximum density. But when he comes to determine the precise place at which that point is found, as it is a matter of experiment, and cannot be got by the mind only, he is at fault; in subsequent experiments learning to become accurate.

He seems to have lowered the point to 36°, and afterwards considered it 38°, the point now apparently fixed on is 39°, or 39.101. (Playfair and Joule.) Dr. Hope's experiments gave it as between 39½ and 40 degrees. In this investigation Dalton's mind again analyses itself, dividing to great clearness of conception on the one side, and carelessness of minute observation on the other.

In 1830 on reading over some old letters which he was arranging, he found one from Dr. Hope, saying, "notwithstanding the caution you gave me, I venture to publish my pamphlet on the contraction of water by heat," Dr. Dalton said, "aye, he had the advantage of me there, but not so much as it appeared at first sight."

In this paper he makes an observation on the power of capillary tubes to prevent the freezing of water, a circumstance which has not been thoroughly inquired into, nor the cause assigned its proper place.

In May, 1800, Dalton was elected secretary of the Literary and Philosophical Society of Manchester, in the place of Dr. William Henry, and having as his colleague Dr. Hull. This office he retained until the year 1808, when he was made vice-president in the room of Dr. Roget, who then lived in Manchester. Soon after, on June 27th, he read to that Society, "Experiments and observations on the heat and cold produced by the mechanical condensation and rarefaction of air." * Here by well devised experiments he endeavoured to shew what however had been before held by Lambert, Saussure, and Pictet, "that the capacity of a vacuum for heat is less than an equal volume of atmospheric air, and that the denser air is, the less is its capacity for heat," indicating a mode of ascertaining "the absolute capacity of a vacuum for heat," and "likewise the capacity of the different gases for heat by a method wholly new." An important result of these experiments was, that the temperature of air mechanically compressed to one-half its volume was raised 50°. This, although much underrated, was the first numerical result of importance on this subject. p. 524.

In this paper we find that he had ascertained that gases expand 1-10th of their volume nearly for 50° of heat, or nearly 1-500th of their bulk, a subject which he treated of at a later period.

Whilst engaged in teaching at the academy in Manchester, his classes or scholars seem to have been as miscellaneous as they had been at Kendal. We may infer this from the appearance of an English grammar, the preface of which is dated March 10th, 1801. He seems to have looked on this as a recreation, but he never afterwards appears to have had recourse to literature for amusement or for variety. As he has never been looked on as a grammarian, it may be of some interest to see what his views on such points were.

He says, p. 8. "It may be taken as an axiom that all time or duration in the strict sense of the terms, is either

* Mem., Vol. V., p. 515.

past or *future.* But for the purposes of speech we must have a present time of some duration, which must necessarily be comprised of a portion of past and a portion of future, having the present, *now* or *instant,* as a boundary between them. Its length may be what we please to make it.

" Grammatically speaking therefore, there are three times, present, past, and future; though strictly and mathematically speaking, we can admit only two, past and future. Moreover we find it expedient in the course of conversation, not only to mention actions as whole and entire, but also their commencement, their being in a passing or middle state, and their termination; accordingly our language furnishes us with four forms of speech for each of the times or tenses, which are exemplified in their proper place, both for the active and passive verb, with appropriate names to them."

His active verb is given thus:—

Indicative mood.

Present time.

I serve, &c.

Beginning present.

I am about to serve, &c.

Middle present.

I am serving, &c.

Ending present.

I have served or been serving, &c.

Past time.

I served, &c.

Beginning past.

I was about to serve, &c.

Middle past.

I was serving, &c.

Ending past.

I had served or been serving, &c.

Future or present time.

I shall, will, may, can *or* must serve, &c.

Beginning future or present.

I shall, will, may, can *or* must be about to serve, &c.

Middle future or present.

I shall, will, may, can *or* must be serving, &c.

Ending future or present.

I shall, will, may, can, *or* must have served *or* have been serving.

In grammar it is difficult to have absolutely new ideas, the subject has been so belaboured, and at the same time it is not easy to keep rigidly to any system proposed, so many of the treatises have wanted clearness. We may see that in that department Dalton was inclined to be an innovator, although he has not the honor of being a discoverer, indeed his mind was much too rigid to be inclined to yield to all the flexions and variations of a subject so bordering on metaphysics as grammar. Horne Tooke is the writer which he most admired on that subject, using sometimes his very words, although not in all things following him. But innovators are more dangerous in grammar, and are less easily received, than in the physical sciences which have no ancestry.

Some years afterwards he went into the shop of the publisher of his grammar, and asked for a copy; he was told they had none, but insisting on it, a parcel of them was found in some dusty corner, very few having ever been sold. Still he assures us that a Sheffield man had published it some years later as his own, with some additions.

In October of the same year he read a paper which occupied three evenings of the Literary and Philosophical Society. It is composed of four "Experimental essays on the constitution of mixed gases; on the force of steam or vapour from water and other liquids in different temperatures, both in a Torricellian vacuum and in air; on evaporation; and on the expansion of gases by heat." (Mem. Vol. V., p. 535.)

The four laws given by him are—

"1. When two elastic fluids, denoted by *A* and *B*, are mixed together, there is no mutual repulsion amongst their particles; that is, the particles of *A* do not repel those of *B*, as they do one another. Consequently, the pressure or whole weight upon any one particle arises solely from those of its own kind.

"2. The force of steam from all liquids is the same, at equal distances above or below the several temperatures at which they boil in the open air; and that force is the same under any pressure of another elastic fluid as it is in vacuo. Thus the force of *aqueous* vapour of 212° is equal to 30 inches of mercury; at 30° below, or 182°, it is of half that force; and at 40° above, or 252°, it is of double the force; so likewise the vapour from sulphuric ether, which boils at 102°, then supporting 30 inches of mercury, at 30° below that temperature it has half the force, and at 40° above it, double the force; and so in other liquids. Moreover the force of aqueous vapour of 60° is nearly equal to $\frac{1}{2}$ inch of mercury, when admitted into a Torricellian vacuum; and water of the same temperature, confined with perfectly dry air, increases the elasticity to just the same amount.

"3. The quantity of any liquid evaporated in the open air is directly as the force of steam from such liquid at its temperature, all other circumstances being the same.

"4. All elastic fluids expand the same quantity by heat; and this expansion is very nearly in the same equable way as that of mercury; at least from 32° to 212°. It seems probable the expansion of each particle of the same fluid, or its sphere of influence, is directly as the quantity of heat combined with it; and consequently the expansion of the fluid as the cube of the temperature, reckoned from the point of total privation."

The first law accounts for a diffusion of gases to a great extent, but not entirely. It would result from it, if not qualified, that there would be a diminishing quantity of oxygen, which is the heaviest gas in the atmosphere, according as the height increased. This was Dalton's opinion, but it has not turned out to be the case. This law was much assailed, and at the same time much misunderstood. The objection that vapour did not rise so rapidly in air as in a vacuum seemed to him a strong one, which he did not quite get over, but considered it as presenting the same difficulty to all theories of the solution of water in air.

The law was stated too broadly, it did not even allow room for the impenetrability of matter to have its due place, and many persons supposed it to mean that a space filled with one gas, might be filled with an equal quantity of another.

He subsequently stated these two propositions in the following form, which he published in the second edition of his "New system of chemistry," when, after many years, he reviewed himself and his reviewers. p. 191, Part I., 1842.

"1. The diffusion of gases through each other is effected by means of the repulsion belonging to the homogeneous particles; or to that principle which is always energetic to produce the dilatation of the gas.

"2. When any two or more mixed gases acquire an equilibrium, the elastic energy of each against the surface of the vessel or of any liquid, is precisely the same as if it were the only gas present occupying the whole space, and all the rest were withdrawn."

There is no doubt that the law had been hastily expressed: explaining some points, it contradicted others. The phenomenon of the mixing of gases is easily explained, if we admit the constant intestine motion of the particles to be a necessary condition of the existence of a body in a gaseous state. (See a paper "On the changes of temperature produced by the rarefaction and condensation of air," by J. P. Joule. Phil. Magaz., May, 1845.)

The second essay is on the force of steam or vapour. He

gives a long table of the force of aqueous vapour at different temperatures, from 40° to 325°. Between 32° and 312° the numbers are given from experiment; above and below these limits the numbers are from calculation. These tables were afterwards modified by himself, and others have also reduced them to greater accuracy. He objects to the tables from water and alcohol given by M. Betancourt in 1790, and to that in the Encyclopædia Britannica, because the authors had assumed the force of that from water, at 32°, to be nothing. This constituted one of the steps which the subject made in its rather retarded progress.

He gives a series of experiments on the power of vapour from liquids, supporting his conclusions by experiments on ether, alcohol, water of ammonia, solution of muriate of lime, mercury, and sulphuric acid, and says "That the variation of the force of vapour from all liquids is the same for the same variation of temperature, reckoning from vapour of any given force; thus assuming a force equal to 30 inches of mercury as the standard, it being the force of vapour from any liquid boiling in the open air, we find aqueous vapour loses half its force by a diminution of 30° of temperature; so does the vapour of any other liquid lose half its force by diminishing its temperature 30° below that in which it boils, and the like for any other increment or decrement of heat." p. 564.

When speaking of vapour of water in air, he says "the results of all agree in one general rule or principle, which is this; let 1 represent the space occupied by any kind of air of a given temperature, and free from moisture; $p =$ the given pressure upon it in inches of mercury; $= f =$ the force of vapour from any liquid in that temperature in vacuo; then the liquid being admitted to the air, an expansion ensues, and the space occupied by the air becomes immediately and in a short time $= 1 + \frac{f}{p-f}$; or which is the same thing $= \frac{p}{p-f}$. Thus in water for instance, let $p = 30$ inches $f = 15$ inches to the given temperature 180°. Then $\frac{p}{p-f} = \frac{30}{30-15} = 2$ for the space; or the air becomes of twice the bulk." p. 572. "In short, in all cases the vapour arises to a certain force according to temperature, and the air adjusts the equilibrium by expanding and contracting as may be required."

"The notion of a chemical affinity subsisting between the gases and vapours of different kinds cannot at all be reconciled to these phenomena." p. 574.

This notion of chemical affinity holding the gases in solution had begun to die out.

In essay third, "*On evaporation*," he concludes that the quantity of any liquid evaporated in the open air is directly as the force of steam from such liquid at its temperature, all other circumstances being the same. He adds a "table shewing the force of vapour, and the full evaporating force of every degree of temperature from 20° to 85°, expressed in grains of water that would be raised per minute from a vessel of six inches in diameter, supposing there were no vapour already in the atmosphere." p. 585. He obtained the evaporation from a surface when the air was still and when in motion. He adds also rules to find the amount of water that can be evaporated from a given surface when the temperature of the air is given, and the condensing point, and to find the force of the aqueous vapour.

The fourth essay on the expansion of elastic fluids by heat proves the law already stated.

The position of the question when he took up the subject may best be explained by himself, he says, p. 595, "The principal occasion of this essay is another on the same subject by Messrs. de Morveau and du Vernois, in the first vol. of the Annales de Chimie. It appearing to them that the results of the experiments of De Luc, Col. Roy, de Saussure, Priestley, Vandermonde, Berthollet, and Monge, did not sufficiently accord with each other; and that it would be of importance to determine not only the whole expansion of each gas from two distant points, such as the freezing and boiling, but likewise whether that expansion be uniform in every part of the scale, they instituted a set of experiments expressly for those purposes. The result of which was; that betwixt the temperatures of 32° and 212°, the whole expansion of one gas differs much from that of another, it being in one instance about 4-10ths of the original, and in others, more than twelve times that expansion; and that the expansion is much more for a given number of degrees in the higher than in the lower part of the scale. These conclusions were so extremely discordant with and even contradictory to those of others, that I could not but suspect some great fallacy in them, and found it in reality to be the fact; I have no doubt it arose from the want of due care to keep the apparatus and materials free from moisture."

After giving his experiments on air, hydrogen, oxygen, carbonic and nitrous gas, in which "the small differences never exceeded six or eight parts, on the whole 345," he adds, "Upon the whole, therefore, I see no sufficient reason why we may not conclude that all elastic fluids under the same pressure expand equally by heat, and that for any given expansion of mercury, the corresponding expansion of air is proportionally something less, the higher the temperature."

"This remarkable fact that all elastic fluids expand the same quantity in the same circumstances, plainly shews that the expansion depends solely upon heat; whereas the expansion in solid and liquid bodies seems to depend on an adjustment of the two opposite forces of heat and chemical affinity, the one a constant force in the same temperature, the other a variable one, according to the nature of the body; hence the unequal expansion of such bodies. It seems, therefore, that general laws respecting the absolute quantity and the nature of heat, are more likely to be derived from elastic fluids than from other substances." There is an admirable clear-sightedness in his short and rapid conclusions. The same law of equal expansion of gases was published six months later by Gay Lussac, and is often called by his name. Dr. Ure says the experiments were made by Gay Lussac with much more care and exactness, but the newest results obtained by Regnault by no means speak so in favour of Gay Lussac. The difference between his results and Dalton's were only trifling. Gay Lussac gave the expansion per degree at 480, Dalton 483, Regnault 491. In this country we have generally used Gay Lussac's for no sufficient reason. On the Continent Dalton has almost been entirely deprived of his merit, and is not even mentioned in connection with it in many French and German works: but such circumstances are unfortunately of constant occurrence. It is difficult to find the reason of this, but it happens so often that our countrymen are quite omitted in their works, that it must in a great measure arise from their neglect of our literature. This certainly must be the cause, as we find that both French and Germans of high name can treat latent heat without even mentioning the name of Black, whose claims are not even disputed; this last occurs even with the very systematic Gmelin. We can readily imagine how some of the other papers of Dalton have been overlooked as merely additions to a subject, whereas he who gives the polish and establishes the law has been allowed the entire credit. They were certainly put within the reach of inquirists, as he says in a letter quoted in Dr. Henry's life of him, p. 50. "My lately published essays on gases, &c., together with the more recent ones read at our society, and of which I gave the result in my late lectures, have drawn the attention of most of the philosophers of Europe. They are busy with them at London, Edinburgh, Paris, and in various parts of Germany, some maintaining one side and some another. The truth will surely out at last." Although not alluding specially to the last mentioned memoir, this letter alludes to his investigations generally, which had been everywhere discussed.

On November 12th, 1802, he read to the Literary and Philosophical Society an "Experimental inquiry into the proportion of the several gases or elastic fluids constituting the atmosphere."*

These he made by weight, p. 257.

Azotic gas	75.55
Oxygenous gas	23.32
Aqueous vapour	1.03
Carbonic acid gas	10
	100.00

In another place we find, by bulk...... 79 azote.
21 oxygen.

In describing his Eudiometric process he has a few observations of great importance, indications of the direction in which he was moving, but given in such a way as to lead us to the conclusion that he had not yet seen their value; teaching us also that an idea of definite proportions may exist without any distinct nature of the completeness of the law of equivalents as it stands. At page 249 he says,—

"2. If 100 measures of common air be put to 36 of pure nitrous gas in a tube 3-10ths of an inch wide and 5 inches long, after a few minutes the whole will be reduced to 79 or 80 measures, and exhibit no signs of either oxygenous or nitrous gas.

"3. If 100 measures of common air be admitted to 72 of nitrous gas in a wide vessel over water, such as to form a thin stratum of air, and an immediate momentary agitation be used, there will, as before, be found 79 or 80 measures of pure azotic gas for a residuum.

"4. If in the last experiment, less than 72 measures of nitrous gas be used, there will be a residuum containing oxy-

* 1st vol. of Memoirs, new series, p. 244.

genous gas; if more, then some residuary nitrous gas will be found.

"These facts clearly point out the theory of the process: the elements of oxygen may combine with a certain portion of nitrous gas, or with twice that portion, but with no intermediate quantity. In the former case *nitric* acid is the result, in the latter *nitrous* acid; but as both these may be formed at the same time, one part of the oxygen going to one of nitrous gas and another to two, the quantity of nitrous gas absorbed should be variable; from 36 to 72 per cent. for common air. This is the principal cause of that diversity which has so much appeared in the results of chemists on this subject."

In the paper on the expansion of elastic fluids, he had already, in a plate, shown that he was accustomed to view gases as composed of definite particles, having drawn each with a different form.

Immediately after this, January 28th, 1803, he read, an inquiry "On the tendency of elastic fluids to diffusion through each other."* This subject was first begun by Priestley. The memoir which he has written on the transmission of gases through porous vessels, entitled "Experiments relating to the seeming conversion of water into air," is certainly one of the most beautiful specimens of investigation that can anywhere be found. He there establishes the fact, that through porous vessels, gases pass one way, vapour of water and other liquids another; and observed, that the mercury in one experiment had risen 3½ inches above the level on the outside. He afterwards found that what could take place with "air and water, will be done with any two kinds of airs."

He failed, however, to make the next step, having said that it is probable "that if two kinds of air of very different specific gravities, were put into the same vessel with very great care, they might continue separate," although his

* Memoirs, Vol. 1., New Series, p. 259.

own experiments justified a different opinion. Dalton took the subject up at this stage, and says the result "establishes this remarkable fact, that a lighter elastic fluid cannot rest upon a heavier, as is the case with liquids; but they are constantly active in diffusing themselves through each other until an equal equilibrium is effected; and that without any regard to their specific gravity, except so far as it accelerates or retards the effect according to circumstances."*

"The only apparatus found necessary, was a few phials and tubes with perforated corks; the tube mostly used was one 10 inches long, and of 1-20th inch bore; in some cases a tube of 30 inches in length, and 1-3rd inch bore was used; the phials held the gases that were subjects of experiment, and the tube formed the connection." p. 261. This tube was often a piece of tobacco pipe.

He believes that this proves his theory of elastic fluids to be correct, that gases are as a vacuum to each other, and it no doubt does favour it, especially as he added that they might be obstructed as a stream of water by a stony bed. Still this very explanation takes away much of the original meaning, and any of his difficulties as to the mutual action of gases must be cleared by further experiments, as has been the case with the laws of diffusion which have already been shewn to us by Professor Graham. There is no doubt that Dalton's expression is an useful attempt to grasp a great difficulty, not yet grasped, we shall see him returning to it again in the next paper.

On October 21st, 1803, he read to the Literary and Philosophical Society, another investigation "On the absorption of gases by water and other liquids." p. 271., Vol. I., New Series.

In this he says, 2. "If a quantity of water freed from air be agitated in any kind of gas not chemically uniting with water, it will absorb its bulk of the gas, or otherwise a

* Page 260, Vol. I., New Series.

part of it, equal to some one of the following fractions, namely, 1-8th, 1-27th, 1-64th, 1-125th, these being the cubes of the reciprocals of the natural numbers 1, 2, 3, &c.;" This has not found general assent, nor can it flow from any known natural law; indeed if it were true it would not shew itself by the usual mode of experimenting, as we can readily imagine one part of the water having 1-4th, another 1-5th, both being distinct parts of the whole, but so mixed with each other in the water that no result is perceived.

4. "If a quantity of water free from air be agitated with a mixture of two or more gases, such as atmospheric air, the water will absorb portions of each gas the same as if they were presented to it separately in their proper density."

5. "If water impregnated with any one gas (as hydrogenous) be agitated with another gas equally absorbable (as azotic), there will apparently be no absorption of the latter gas; just as much gas being found after agitation as was introduced to the water; but upon examination the residuary gas will be found a mixture of the two, and the parts of each, in the water, will be exactly proportional to those out of the water."

"10. Pure distilled water, rain and spring water, contain nearly their due share of atmospheric air; if not, they quickly acquire that share by agitation in it, and lose any other gas they may be impregnated with. It is remarkable however that water by stagnation in certain circumstances loses part or all of its oxygen, notwithstanding its constant exposition to the atmosphere. This I have uniformly found to be the case in my large wooden pneumatic trough, containing about 8 gallons. * * * * The quantity of azotic gas is not materially diminished by stagnation, if at all." He has not here considered the action of the organic substances.

Theory of the absorption of gases by water. p. 283.

"1. All gases that enter into water and other liquids, by means of pressure, and are wholly disengaged again by the removal of that pressure, are mechanically mixed with the liquid, and not chemically combined with it."

He had already mentioned Dr. Henry's discovery, that the quantity of gas absorbed is as the density or pressure.

"2. Gases so mixed with water, &c., retain their elasticity or repulsive power amongst their own particles, just the same in the water as out of it, the intervening water having no other influence in this respect than a mere vacuum."

"3. Each gas is retained in water by the pressure of gas of its own kind incumbent on its surface abstractedly considered, no other gas with which it may be mixed having any permanent influence in this respect."

"4. When water has absorbed its bulk of carbonic acid gas, &c., the gas does not press on the water at all, but presses on the containing vessel just as if no water were in. When water has absorbed its proper quantity of oxygenous gas, &c., that is, 1-27th of its bulk, the exterior gas presses on the surface of the water with 26-27ths of its force, and on the internal gas with 1-27th of its force, which force presses upon the containing vessel, and not on the water. With azotic and hydrogenous gas the proportions are 63-64ths and 1-64th respectively. When water contains no gas, its surface must support the whole pressure of any gas admitted to it, till the gas has in part forced its way into the water."

"5. A particle of gas pressing on the surface of water is analogous to a single shot pressing upon the summit of a square pile of them. As the shot distributes its pressure equally amongst all the individuals forming the lowest stratum of the pile, so the particle of gas distributes its pressure equally amongst every successive horizontal stratum of particles of water downwards, till it reaches the sphere of influence of another particle of gas. For instance, let any gas press with a given force on the surface of water, and let the distance of the particles of gas from each other be to those of water as 10 to 1, then each particle of gas must divide its force equally amongst 100 particles of water, as follows: It exerts its immediate force upon 4 particles of water; those 4 press upon 9, the 9 upon 16, and so on according to the order of square numbers, till 100 particles of water have the force distributed amongst them; and in the same stratum each square of 100, having its incumbent particle of gas, the water below this stratum is uniformly pressed by the gas, and consequently has not its equilibrium disturbed by that pressure."

"When water has absorbed 1-27th of its bulk of any gas, the stratum of gas on the surface of the water presses with 26-27ths of its force on the water, and with 1-27th of its force on the uppermost stratum of gas in the water; the distance of the two strata of gas must be nearly 27 times the distance of the particles in the incumbent atmosphere, and 9 times the distance of the particles in the water. This comparatively great distance of the inner and outer atmosphere arises from the great repulsive power of the latter, on account of its superior density, or its presenting 9 particles of surface to the other 1. When 1-64th is absorbed, the distance of the atmospheres becomes 64 times the distance of two particles in the outer, or 16 times that of the inner. The annexed views of perpendicular and horizontal strata of gas in and out of water will sufficiently illustrate these positions."*

"7. An equilibrium between the outer and inner atmospheres can be established in no other circumstance than that of the distance of the particles of one atmosphere being the same or some multiple of that of the other; and it is probable the multiple cannot be more than 4. For in this case the distance of the inner and outer atmospheres is such as to make the perpendicular force of each particle of the former or those particles of the latter that are immediately subject to its influ-

* A plate accompanied this.

ence, physically speaking, equal; and the same may be observed of the small lateral force."

"8. The greatest difficulty attending the mechanical hypothesis arises from different gases observing different laws. Why does water not admit its bulk of every kind of gas alike? This question I have duly considered, and although I am not yet able to satisfy myself completely, I am nearly persuaded that the circumstance depends upon the weight and number of the ultimate particles of the several gases; those whose particles are lightest and single being least absorbable, and the others more, according as they increase in weight and complexity—(subsequent inquiry made him think this less probable). An inquiry into the relative weights of the ultimate particles of bodies is a subject, as far as I know, entirely new; I have lately been prosecuting this inquiry with remarkable success. The principle cannot be entered upon in this paper; but I shall just subjoin the results, as far as they appear to be ascertained by my experiments."

He then gives a list of *relative weights* of 21 substances, constituting the first attempt to form a table of atomic weights.

"Table of the relative weights of the ultimate particles of gaseous and other bodies.

Hydrogen	1
Azot	4.2
Carbone	4.3
Ammonia	5.2
Oxygen	5.5
Water	6.5
Phosphorus	7.2
Phosphuretted hydrogen	8.2
Nitrous gas	9.3
Ether	9.6
Gaseous oxide of carbone	9.8

Nitrous oxide	13.7
Sulphur	14.4
Nitric acid	15.2
Sulphuretted hydrogen	15.4
Carbonic acid	15.3
Alcohol	15.1
Sulphureous acid	19.9
Sulphuric acid	25.4
Carburetted hydrogen from stagnant water	6.3
Olefiant gas	5.3"

I have given as much as possible, in his own words, the most important points attended to by Dalton up to this date. It was not my intention to inquire into the particulars relating to the novelty of the views taken by him, except on the atomic theory, and have therefore purposely left out any such opinions as might require discussion; nor have I shewn in all cases where advancing science has differed from his results. Some things in the papers alluded to were bold and strikingly new, some things are improvements on the old, some are mere re-statements of the old, but all is done in a firm, clear, and determined manner, as by a master in the business, going to the real point of difficulty in every case, and at all times avoiding unimportant details or vain ornament. He drives on like a new settler, and clears the ground before him, leaving it rather rugged it is true; nevertheless it is resolutely cleared.

CHAPTER III.

DALTON'S SOCIAL LIFE.

We now approach the most important discovery of Dalton, and before entering upon it, it may be well to acquaint the reader with the general character and appearance of the man in his vigour. It is not my intention, as before stated, to amuse myself or readers with many little incidents of his life, nor can we gain by looking at such a character apart from the student of nature; but it is needful to give some slight account of the appearance and habits of the agent by which such valuable knowledge of natural law has been gained. On his habits, Miss Johns's Journal, lent me freely by Mr. Woolley, is the best authority. The Rev. W. Johns, once a colleague of Dalton's at the academy, had a school in George-street, near the Literary and Philosophical Society, which had given up a portion of its room to Dalton. In the autumn of 1804, Mrs. Johns saw him casually pass, and asked him why he never came to see them. Dalton said, "I do not know; but I will come and live with you, if you will let me." He did so, and took possession of the only bedroom at liberty, sitting with the family. In this family he lived for twenty-six years in the greatest amity, until Mr. Johns, giving up the school, sought a purer air in the suburbs of the town.

The portrait which is appended to this memoir, is from a picture by Allen, presented to the society by that painter on the occasion of Dalton being made president. It represents him in the vigour of life, and must of course be a more suitable representation of the man than those taken in old age, although one at least of those by Stephenson is an excellent portrait of a late period, and a beautiful engraving.

He was rather above the middle size, five feet seven inches. Mr. Giles, who read a memoir of him to the Manchester Society, and who was his pupil, says that "he was robust, athletic, muscular, and stooped slightly as if hasting forward, for he was a rapid walker. His countenance was open and manly. His voice was deep and gruff; and his lectures were by no means interesting, except to those who were satisfied with matter independent of style: he even spoke in a careless and mumbling manner." He was a Quaker, as has been said, and dressed in their peculiar manner, taking care that every article of dress should be of the finest texture, but avoiding the extreme of formality. In his general conversation he did not adopt their style; and never gave any opinion on religious subjects. His most intimate friends found him reserved on such points, although at times they found that there was in him great reverence and deep feeling. But he evidently did not think much on religious subjects, and seems not even to have formed strong opinions upon them, giving way to the opinions of those around him, like one unable or unwilling to form them for himself. If this were the case only with such subjects as are peculiarly religious, we might suppose that it arose simply from a want of agreement with current opinions which he was unwilling to disturb, but as he stood in the same attitude towards metaphysical opinions, we may fairly conclude that those faculties which discuss the moral and intellectual history and position of man, were not highly developed in Dalton. It would not be wise to conclude that they were weak, because we find that he had a great power of concentration, and in his ardent study of a subject he seems to have allowed the rest of his mind to be satisfied with meditative culture. In early life he seemed inclined to answer moral and metaphysical questions in the *Gentleman's Magazine;* and occasionally a lady would induce him to write a few lines of such poetry as a well-educated man is generally able to write. But his life was spent in his laboratory, and all subjects not connected with his pursuits were much neglected; he might have said of their cultivation, as he said when asked why he never got married, "I never had time."

He rose at about eight o'clock in the morning; if in winter, went with his lantern in his hand to his laboratory, lighted the fire, and came over to breakfast, when the family had nearly done. Went to the laboratory and staid till dinner time, coming in a hurry when it was nearly over, eating moderately and drinking water only. Went out again and returned at about five o'clock to tea, still in a hurry, when the rest were finishing. Again to his laboratory till nine o'clock when he returned to supper, after which he and Mr. Johns smoked a pipe, and the whole family seems much to have enjoyed this time of conversation and recreation after the busy day. Dr. Schunck, who dined there as a child at school, says he never appeared in a hurry.

He was rather silent, especially if the company were large, but an attentive listener, whilst he occasionally introduced some short sentence of dry humour. With a few of his intimate friends he enjoyed much a lively conversation, but does not seem to have been fitted for dealing with men assembled in large numbers, either in public or private. This did not arise from any want of self-possession, which is said never to have been known to be ruffled: an illustration of this is given in the following. When in using the air-pump at a lecture a glass vessel burst, making a considerable noise and causing the ladies to scream, he simply said, "that is more than I intended, it's broken," and went on again. His disinclination to speak made him, as a teacher, by no means communicative; he allowed his pupils to learn, and willingly answered a question, but during the most of the time he was attending to his experiments, thinking, probably, that they were much better off than he ever was to have some one to apply to if a difficulty arose.

His habits were careful and economical, some say parsimonious, but he was by no means wanting in generosity, and gave fifty pounds to the building of the new Meeting House, at a time when he certainly could have had but very little. Such men do not often seek amusements, and he had only one, a remnant of Cumberland, which he seemed never to forget, at the same time also a wholesome exercise. Every Thursday afternoon, about two o'clock, he went outside the town to the "Dog and Partridge," now far within Manchester, and played a few games at bowls. This he seems to have thoroughly enjoyed, watching the bowls with the greatest anxiety, and by his constant movements indicating, as people are apt to do, the way in which they wish the bowl to move, as if endeavouring to influence it. He shewed even there a glimmer of the latent enthusiasm of his mind. He played a fixed number of games and then ceased, took tea at the inn, smoked his pipe, and went to his laboratory. Between twelve and one he usually went to the Portico to read the newspapers, but did not strongly speak on political subjects, so that even the Johns family did not for a long time know that he was Conservative in politics. At the same time Mr. Giles says he was a Liberal, always voting for Liberals, so that we may call him a liberal Conservative, which however indefinite, is somewhat the character of such a man, considerate, desirous of improvement, but not inclined to violent change.

His great delight was to visit the hills of Cumberland where he first studied the clouds and the aurora, and when the usual day of June had come, old Matthew Jobson came out of his cottage under the slopes of Helvellyn, and looked out for Dalton and his instruments. He ascended the hill from thirty to forty times during his life, walking rapidly and with ease, generally keeping before any party who accompanied him, so as on one occasion to have brought out the exclamation from a friend, "John, I wonder what thy legs are made of." In later life occasionally he was accompanied by some of the younger members of the family to whom he was now as a relation, and we can well believe Miss Johns, when she says, "that to those who have seen him only on ordinary occasions, it is impossible to convey an idea of his enthusiasm on those occasions. He never wearied." Jonathan Otley, the veteran guide, at Keswick, who has spent his life mapping, describing, and showing the country around, often accompanied Dalton, and he has in a journal given an account of some of their excursions. These were undertaken partly with a view of "bringing into exercise a set of muscles, which would otherwise have grown stiff," as Dalton said, and partly to make meteorological observations, or to bring down air for analysis from the highest points of the county, gas from the floating island, and minerals from every hill.

In order to give his habits of thought on objects not scientific, a few of his letters may be introduced with advantage; at the same time they will give an account of some portions of his life, which would lose much of their interest if the words of another were alone used.

Having been invited to lecture at the Royal Institution, he thus describes his visit in a letter to his brother, February 1st, 1804. His first visit to London had been made in 1792.

"Dear Brother, I have the satisfaction to inform thee that I returned safe from my London journey, last seventh day, having been absent six weeks. It has, on many accounts, been an interesting vacation to me, though a laborious one. I went in a great measure unprepared, not knowing the nature and manner of the lectures in the institution, nor the apparatus. My first was on Thursday, December 22nd (1803), which was introductory, being entirely written, giving an account of what was intended to be done, and natural philosophy in general. All lectures were to be one hour each, or as near as might be. The number attending were from one to three hundred of both sexes, usually more than half men. I was agreeably disappointed to find so learned and attentive an audience, though many of them of rank. It required great labour on my part to get acquainted with the apparatus and to draw up the order of experiments and repeat them in the intervals between the lectures, though I had one pretty expert to assist me. We had the good fortune, however, never to fail in any experiment, though I was once so ill prepared as to beg the indulgence of the audience, as to part of the lecture, which they most handsomely and immediately granted me by a general plaudit. The scientific part of the audience was wonderfully taken with some of my original notices relative to heat, the gases, &c., some of which had not before been published. Had my hearers been generally of the description I had apprehended, the most interesting lectures I had to give, would have been the least relished, but as it happened, the expectations formed had drawn several gentlemen of first-rate talents together; and my eighteenth, on heat, and the cause of expansion, &c., was received with the greatest applause, with very few experiments. The one that followed was on *mixed elastic fluids*, in which I had an opportunity of developing my ideas, that have already been published on the subject more fully. The doctrine has, as I apprehended it would, excited the attention of philosophers throughout Europe. Two journals in the German language, came into the Royal Institution, whilst I was there, from Saxony, both of which were about half filled with translations of the papers I have written on the subject, and comments on them. Dr. Ainslie was occasionally one of my audience, and his sons constantly: he came up at the concluding lecture, expressed his high satisfaction, and he believed it was the same sentiment with all or most of the audience. I was at the Royal Society one evening, and at Sir Joseph Banks's another. This gentleman I had not, however, the pleasure of seeing, he being indisposed all the time I was in London.

"I saw my successor, William Allen, fairly launched; he

gave his first lecture on Tuesday preceding my conclusion. I was an auditor in this case, the first time, and had an opportunity of surveying the audience. Amongst others of distinction the Bishop of Durham was present.

"In lecturing on optics I got six ribbands, blue, pink, lilac, and red, green, and brown, which matched very well and told the curious audience so. I do not know whether they generally believed me to be serious, but one gentleman came up immediately after and told me he perfectly agreed with me: he had not remarked the difference by candle light."

This letter concludes characteristically by "The rain has been 27½ inches last year."

Like many students whose nerves are not easily affected, Dalton liked tobacco. A thorough explanation of its action on various constitutions seems to have hitherto escaped the research of medical men, most of them being content either to admire it, or to detest it, according as it may suit themselves. The following letter to Mr. John Rothwell gives Dalton's taste.

London, Jan. 10th, 1804.

"I was introduced to Mr. Davy, who has rooms adjoining mine in the Royal Institution: he is a very agreeable and intelligent young man, and we have interesting conversations in an evening: the principal failing in his character is, that he does not smoke. Mr. Davy advised me to labour my first lecture: he told me the people here would be inclined to form their opinion from it; accordingly I resolved to write my first lecture wholly; to do nothing but to tell them what I would do, and enlarge on the importance and utility of science. I studied and wrote for nearly two days, then calculated to a minute how long it would take me reading, endeavouring to make my discourse about fifty minutes. The evening before the lecture, Davy and I went into the theatre: he made me read the whole of it, and he went into the furthest corner; then he read it, and I was the audience; we criticised upon each other's method. Next day I read it to an audience of about 150 or 200 people, which was more than were expected. They gave a very general plaudit at the conclusion, and several came up to compliment me on the excellence of the introductory. Since that time I have scarcely written anything; all has been experiment and verbal explanation. In general my experiments have uniformly succeeded, and I have never once faltered in the elucidation of them. In fact, I can now enter the lecture room with as little emotion nearly as I can smoke a pipe with you on Sunday or Wednesday evenings."

In 1807 he gave a similar course of lectures in Edinburgh, on which the following addressed to his friend, the Rev. W. Jones, may be read with interest.

Edinburgh, April 19th, 1807.

"Respected Friend,

As the time I proposed to be absent is nearly expired, and as my views have recently been somewhat extended, I think it expedient to write you for the information of enquirers. Soon after my arrival here I announced my intention by advertisement of handbills; I obtained introduction to most of the professional gentlemen in connection with the college, and to others not in that connection, by all of whom I have been treated with the utmost civility and attention; a class of eighty appeared for me in a few days; my five lectures occupied me nearly two weeks; they were finished last Thursday, and I was preparing to leave the place, and return by Glasgow, to spend a week. But several of the gentlemen who had attended the course represented to me that many had been disappointed in not having been informed in time of my intention to deliver a course, and that a number of those who had attended the first course would be disposed to attend a second. I have been induced to advertise for a second, which, if it succeeds, will commence on Wednesday, the 22nd, and be continued daily, till the conclusion. This will detain me a week yet; I then set off for Glasgow, where I may be detained for a week or more, so that I see no probability of reaching Manchester before the beginning of May, to which I look forward with some anxiety. Hitherto I have been most highly gratified with my journey; it is worth coming 100 miles merely to see Edinburgh. It is the most romantic place and situation I ever saw; the houses touch the clouds; at this moment I am as high above the ground as the cross on St. James's spire; yet there is a family or two above me; in this place they do not build houses side by side as with you, they build them one upon another, nay, they do what is more wonderful still, they build one street upon another; so that we may in many places see a street with the people in it, directly under one's feet, at the same time that one's own street seems perfectly level and to coincide with the surface of the earth. My own lodgings are up four flights of stairs from the front street, and five from the back. I have just 100 steps to descend before I reach the real earth. I have a most extensive view of the sea; at this moment I see two ships; and mountains across the Firth of Forth, at the distance of thirty miles; to look down from my windows into the street at first made me shudder, but I am now got so familiar with the view, that I can throw up the window and rest on the wall, taking care to keep one foot as far back in the room as I can, to guard the centre of gravity. The walks about Edinburgh are most delightfully romantic. The weather is cold; ice every morning, and we had a thick snow a few days ago. Upon walking up on to an eminence I observed all the distant hills white; the nearer ones speckled; I saw five or six vessels just touching the horizon; they seemed to be about ten or twelve miles off, and their white sails looked like specks of snow on the sea. I saw a dozen or two at anchor in the river, and a most charming view of the Fifeshire hills on the other side of the Firth. Adieu. My best regards to you all."

J. DALTON.

Again, from London, December 27th, 1809, when giving another course of lectures, he writes to Mr. Jones, after some pleasant gossip about his fellow-travellers:—

"On Tuesday I spent greater part of the day (morning they call it here) with Mr. Davy in the laboratory of the Royal Institution. Sir I. Sebright, M.P., who is becoming a student of chemistry, was present. We had a long discussion. In the evening I walked three miles into the city, to Pickford's, to look after my boxes; I found them there, but as they promised to send them next day I did not take them. They disappointed me. On Wednesday I attended Mr. Bond's lecture on astronomy, and prepared for mine the next day. On Thursday, at two, I gave my first lecture. Mr. Pearson, a former acquaintance, went home with me after the lecture, and we had a long discussion on mechanics. Mr. Davy had invited me to dine with the club of the Royal Society, at the Crown and Anchor, at five o'clock, but I was detained till near six; I got there and called Davy out; all was over; the cheese was come out. I went, therefore, to the nearest eating-house I could find to seek a dinner; looking in at a window I saw a great heap of pewter plates and some small oblong tables covered with cloths. I went in and asked for a beefsteak; "no." What can I have? "boiled beef." Bring some immediately. There was nothing eatable visible in the room, but in three minutes I had placed before me a large pewter plate covered completely with a slice of excellent boiled beef swimming in gravy, two or three potatoes, bread, mustard, and a pint of porter. Never got a better dinner. It cost me 11½d. I should have paid 7s. at the Crown and Anchor. I then went to the Royal Society and heard a summary of Davy's paper on chemistry, and one of Home's on the poison of the rattlesnake: Sir J. Banks in the chair. Davy is coming very fast into my views on chemical subjects. On Friday I was preparing for my second lecture. I received a visit from Dr. Roget. On the evening I was attacked with sore throat. I sweated it well in the night with cloathing, but it was bad on Saturday, and I was obliged to beg a little indulgence of my auditors on the score of exertion. However, I got through better than I expected. I kept in on Sunday and Monday and got pretty well recruited. On Tuesday I had my third lecture, after which I went to dine at a tavern to meet the chemical club. There were five of us, two of whom were Wollaston and Davy, secretaries of the Royal Society; we had much discussion on chemicals. Wollaston is one of the cleverest men I have yet seen here. To-day, that is Thursday (for I have had this letter two or three days in hand), I had my fourth lecture. I find several ingenious and inquisitive people of the audience. I held a long conversation to-day with a lady on the subject of rain-gauges. Several have been wonderfully struck with Mr. Ewart's doctrine of mechanical force. I believe it will soon become a prevalent doctrine. I should tell Mrs. J. something of the fashions here, but it is so much out of my province, that I feel rather awkward. I see the belles of New Bond-street every day, but I am more taken up with their faces than their dresses. I think blue and red are the favourite colours. Some of the ladies seem to have their dresses as tight round them as a drum, others throw them round them like a blanket. I do not know how it happens, but I fancy pretty women look well any how.

I am very regular with my breakfast, but other meals are so uncertain that I never know when or what. Hitherto I have dined at from two to seven o'clock; as for tea I generally have a cup between nine and ten, and, of course, no supper. I am not very fond of this way of proceeding. They say things naturally find their level, but I do not think it is the case in London. I sent for a basin of soup the other day before I went to lecture, thinking I should have a good threepenny worth, but I found they charged me one shilling and ninepence for a pint, which was not better than some of our

Mary's broth. Of course, I could not digest much more of the soup."

Again, the year after,

London, Jan. 29th, 1810.

"You may perhaps have heard from Dr. Henry that I have been nearly as ill as formerly, that I have been nearly poisoned since I came here. I had been about three weeks when I discovered it was the porter which produced the effects.* I have not had a drop since, and have never had any more of the symptoms.

"I have had a pretty arduous work, as you may imagine, having had three lectures to prepare each week; to attend two others, and to visit and to receive visits occasionally besides. I find myself just now in the focus of the great and learned of the metropolis. On Saturday evening I had a discussion with Dr. Wollaston, and a party at Mr. Lowry's. On Sunday evening, last night, I was introduced to Sir Joseph Banks, at his house, by Sir John Sebright. Sir Joseph said, 'Oh, Mr. Dalton, I know him very well; glad to see you; hope you are well, &c.' There were forty or more of the leading scientific characters present, many of whom were my previous acquaintance, such as Sir Charles Blagden, Drs. Wollaston, Marcet, Berger, and Roget; Messrs. Cavendish, Davy, Tennant, Lawson, &c.; we had conversation for about an hour or more in Sir Joseph's library, when the company dispersed. To judge from the number of carriages at the door, it might be a court levee.

"I paid a visit, in company with Dr. Lowry, to Dr. Rees, the other day; we spent an hour in conversation in the doctor's library. The doctor seems a worthy philosopher of the old school; his evening lucubrations are duly scented with genuine Virginia."

* Lead had been found in it. This was probably owing to the use of lead pumps, a very common and dangerous custom, whether used as is commonly the case in public-houses, at least in this neighbourhood, to pump the malt liquor from the cellar, or for water for domestic supply.

Sir Humphrey Davy's opinion of Dalton, given in Dr. Henry's recent life, seems to contain rather harsh and unpleasant expressions, and I think scarcely fair suggestions. He could not have known the man otherwise than externally; nor does he seem to have known well the history of his discoveries. I shall allude to it more when speaking of Dalton as a philosopher; as to the man I only collect the opinions of others, and give a few examples of his character from his letters and actions. His brother, Dr. Davy, also says, "Mr. Dalton's aspect and manner were repulsive. There was no gracefulness belonging to him. His voice was harsh and brawling; his gait stiff and awkward; his style of writing and conversation dry and almost crabbed. In person he was tall, bony, and slender. (1809-10.) He never could learn to swim; on investigating this circumstance he found that his specific gravity was greater than that of water, and he mentioned this in his lectures on natural philosophy in illustration of the capability of different persons for attaining the art of swimming." But he adds, "independence and simplicity of manner and originality were his best qualities. Though in comparatively humble circumstances he maintained the dignity of the philosophical character." So many "best" qualities are seldom found in one man.

This word *brawling* was unintelligible to me until Dr. Schunck suggested *drawling*, as the true meaning. Brawling is unintelligible in connection with such a retiring man. Dr. Schunck says, "no one who saw him could call his appearance repulsive. I recollect him from my childhood and never saw it; and children are very susceptible to repulsive appearance in people. He was, I think, good looking; only his deep set eyes were against him." At any rate, there is no connection between brawling and Dalton. When I saw him his mind was quite broken down; but I agree with my friend, Dr. Schunck, that he was agreeable to look upon.

A writer in the *Quarterly Review*, Vol. XCVI., who had heard him lecture, gives him an unfavourable manner, saying "his voice was harsh, indistinct, and unemphatical, and he was singularly wanting in the language and power of illustration, needful to a lecturer on these high matters of philosophy, and by which Davy and Faraday have given such lustre to their discoveries. Among other instances of his odd appropriation of epithets, we recollect that in treating of oxygen, hydrogen, nitrogen, &c., those great elements which pervade all nature, he generally spoke of them as '*these articles*,' describing their qualities with far less earnestness than a London linen draper would shew in commending the very different *articles* which lie on his shelves."

As to his style of writing, it is before us to judge. These letters shew nothing crabbed. The specimens of scientific writing given are equally free from such blame; and although there are different styles in his own works, I consider that generally his scientific writing is agreeable to read, nor is it by any means more "dry" than the average scientific memoirs.

Still we must consider that his appearance or manner was much against him in the eyes of some persons, the evidence being so strong, but the following letter will show that he was not of a repulsive cast of mind. These letters, from which I quote, were written home to Mr. Johns for himself and the family to read, as may be supposed, and shew, instead of a repulsive, an exceedingly amiable disposition. He was lecturing at Birmingham, and he writes:—

March 17th, 1825.

"We left the Bridgewater Arms, three middle sized and middle aged gentlemen in a small coach; there was room for a thin lady opposite my left hand friend, and I was beginning to think he might have the advantage of me in two respects. We drove down to the Palace Inn, and there I saw a corpulent old gentleman set off towards us much in the shape of a sack of malt; he came up in as straight a line as he well could, and having partly entered, my companion called out 'oh!' with a long continuance. His toes were unfortunately the sufferers. The old man had not yet got settled in his contracted berth before he informed us that he thought the coach a very small one. Indeed, his hat nearly touched the roof when sitting. At Stockport, changed horses, a famous lusty woman having brought them out, her husband (I suppose) having been up late the night before. The old gentleman told us a friend of his travelling in Wales having occasion to call at an inn to breakfast, found the housemaid busy rubbing the irons; she left them, and took his horse to stable. Where is the hostler, said he, that was here the last time I was this way? 'I was him,' said the girl. The landlord told his guest he never had a better hostler in his life than she was. We got near to Macclesfield, when my opposite companion, who did not know England as well as Ann says she knows Europe, asked his fellow-traveller whether we were in Staffordshire or Warwickshire. * * * At Newcastle our geographer remarked that there were two Newcastles at a great distance from each other; 'Newcastle-upon-Tyne,' says he, 'is that in England or in Scotland?' After a short pause, the maltster says, 'it is in Cumberland, I think.' 'No. In Northumberland,' said my left hand friend.

I have been here (Birmingham), at one of the show rooms, to see the new lions, since I was last here. There are many good things; amongst other things, they have got new banisters, for stairs. What is it. Mahogany? No. Lignum vitæ? No. Ebony? No. No sort of wood, nor metal either, it is glass, cut glass."

London, May 15th, 1825.

"I agreed to go with Mr. Lowe, one of the superintendents of the gas works, whom I was previously acquainted with, to spend the night. It is at Highgate, four miles on the North-road; we took a stage at nine, and arrived in about forty minutes; after walking up a steep hill, we came to a row of most delightfully situated houses, on an eminence, looking down on the city and the adjacent country. Whether by night or by day it is delightful; at night, we see the lights of the city forming a circle around us; the general effect is most curious. A darkish cloud hung over the whole foreground; the lights from London illuminated that part over it, whilst on this side it was dark, and on the other side dark, so as to give the appearance of a luminous cloud, interposed between two black clouds. The night was still and fine, and Mr. L. promised me I should hear the nightingale; for one or two sung every night from a grove, 100 yards below his house. We smoked our segars till twelve, then looked out and listened; all silent, no nightingale. We went to bed about three, I was awoke by the melodious song of the nightingale, which continued, without any interruption, till four o'clock. Soon after three I also heard the second and third best singing birds that we have, according to Mr. Blackwall, and the first, or nightingale, all singing together. My bedroom window fronted the S.E. In the morning, at eight, I took hold of the blind string with one hand, and put my other to the opposite side to help it up, as usual, with most others; but to my surprise it spun up to the top of the window of its own accord, in a moment, and such a view as I never witnessed presented itself, except at St. Cloud, near Paris. For half a mile before me rose the tops of trees, from a beautiful grove, belonging to Mrs. Coutts, her house chimneys popping up on the right; over the tops of the trees the country presented itself, interspersed with houses, and beyond, London with its spires; St. Paul's dome, like Helvellyn, right in the front; the river, the Kent hills over London, &c., &c., the sun shining clear, the birds singing, &c., &c. Quite at the foot of my window, a kind of verandah, entwined with shrubs below, about 100 yards square of descending ground, so covered with shrubs as to hold two or three shady seats, and some large shrubs, looking like a wilderness, and in the middle a little plot of green ground and a knot of flowers. In short, it was multum in parvo with a witness. This is the very house occupied by the late Mr. Joyce."

To Mr. Johns, when absent from Manchester.

July 2nd, 1825.

"Yesterday I dined at Dr. Henry's, meeting Professor Almroth, of Stockholm, and a party, together with his own young family and Miss Bailey. Professor Almroth breakfasted with me this morning. He is well read in Shakespere, Sir Walter Scott, and the German literature, as well as in *Shimistry.*

"You will have a new chapel to go to when you come here. The great water hole opposite has now a chapel on it, for Baptists, they say. * * * * I am nearly ready for a jaunt, but whether north or west I do not know till my stick falls."

We see from these letters that he was accustomed to lecture occasionally in all parts of the country, when invited, and we see also enough to let us judge a little of his temper. They are full of very simple kindly feeling, the very act of writing home so many details betokens a disposition to please, and entirely precludes every accusation of vanity or that absurd appearance of separation from his fellows, which we find so frequently the production of bigotry and of ignorance, but which we too often call by the name of dignity. We find him exceedingly pleased with the attentions of scientific men; he had, of course, frequent visits from such as came to Manchester, and foreigners of distinction seem to have pleased him as much as the young members of the family where he resided. At the same time he was tenacious of his opinions, and we find that in describing his conversations with men of science, he generally calls them *discussions,* and generally on scientific subjects we find him too much standing up for his own rights, as his results appeared to him to be. This arose from a peculiarity of mind which seems to have been too much developed in him, as in those also where observation and experience are so tenaciously remembered as to become the only groundwork of opinion, and where the arguments of others are as mere fiction, having no influence upon the reasoning powers. Acting in this spirit, he discouraged reading, and prevented the Literary and Philosophical Society from obtaining a sufficient supply of books. He said, "I could carry all the books I have ever read on my back." * In this, he was evidently forgetting how diverse were the faculties of mankind, and acting also in ignorance of the fact, that he was himself suffering from a want of reading, although it is probably true that he was a gainer in another direction from the same cause. But in summing up features of characters, we find great antagonism in our results, so that we are in error when we make up a very uniform design from what is too often a hastily prepared patchwork, and known to be so by the owners themselves, but which they are prevented from completing by circumstances, if not by time, which limits the progress of all. He, in fact, is often the great man who allows himself to act onesidedly, not for his own pleasure or profit, but because the struggle which he has to maintain, needs all energies to be concentrated on one point of attack.

In this way we may view Dalton. We must see him also as a man having limited assistance from the knowledge of others, ignorant of many of the elegant and easy methods of procuring knowledge and illustrating facts which have become the common inheritance of universities, and accustomed to the society of few only who had similar studies; and when meeting with his contemporaries from capital cities, forgetting that they were not like him, completely immersed in the study of nature, but were also expected to cultivate the soireé and the dinner table. Still sociality was not entirely checked

* Mr. Woolley is my authority.

in Dalton, although it was under subjection to routine. This was seen in the regular manner in which he went on Sunday to dine, at Mayfield, with Mr. Neild, even if the host were not at home, and in the regular manner in which he joined the family of Mr. Johns every evening, enjoying the society of the younger members who had grown up from childhood under his eye, and under the same roof, and in the great pleasure he had in taking them with him in his summer excursions. It was a mild enjoyment, allowing of little enthusiasm, this was reserved for the sterner aspects of nature on the summits of Scawfell or Helvellyn; but it was his nature to be calm, a violent life does not suit the inquirer into nature.

He was simple in his habits by nature and by education, but still more so from his pursuits. Such are always found to be disturbed when wealth, by enlarging the establishment, claims too much the care of the possessor, and when work, ceasing to be urgent, allows us to imagine that the cultivation of an acquaintance is the great business of our life. To the courtly especially, he seemed morose, but that it was merely a question of form, and not of inward feeling, we see additional proof in the letter where he makes a similar complaint of Sir H. Davy, as lacking that geniality which he jocularly represents under the symbol of tobacco.

To a certain extent he was separated from society by the constancy of his work. Many have been separated, and separated themselves for idleness, but for work, few; and whilst the world is overflowing with those who would willingly give themselves to the pleasure of social intercourse, an occasional exception for a higher purpose stands forth as a subject for our admiration, surely not for a censure.

Dalton never married, he had not time, he said. In early life his position prevented him, in middle life constant work, when he seems to have been struggling for an independency, not knowing that it would come to him with ease as soon as his failing strength demanded it. This desire to become independent of his work induced in him an undue amount of care in the accumulation of his savings: we may consider them as the representatives of a certain amount of time entirely lost to all but his heirs. At the same it is to be remembered that he was accumulating his savings for a time, when he could no longer be able to work, and the simplicity and self-denial of such a course, instead of being worthy of blame, is a virtue, which, unfortunately, is not of sufficient occurrence. This virtue gave him the opportunity of showing kindness to many of his friends, and of helping such as were in need.

He was on terms of friendship with several ladies whom he greatly admired, and there is little doubt that in one case in early life the admiration was that of love. Whether the fact of the lady's engagement to another affected him for any length of time with disappointment, is what his reserved nature never told to any one, but we are left to guess that the attachment was strong, when late in life he could not read without emotion, and even tears, some verses the lady had written, or allow any one else to read them in the letter. Tears and emotion were rare with him, and leave us room enough for speculating on what might have been the inner romance of that life which externally seemed so simple and so contented with matters of fact.

His attention to ladies, and his great respect for their mental attainments, makes us still more inclined to refer to awkwardness of manner, the appearance of "repulsiveness and harshness," words that seem out of place when used in speaking of one so little disposed to offend.

The answer to his friend Mr. Gough, who attacked him on the subject of the atmosphere, shows great forbearance and innate nobleness of feeling, under circumstances in which every thing that is most bitter and severe is generally admitted to pardon.

In a town like Manchester, where exertions to extend the

material portion of civilization take the form in the minds of most men of a struggle for wealth, in which the original object is forgotten, it is a proof of simplicity and singleness of character when we find that he was never once led away by the glare of the princely fortunes around him. He gave lessons for very small fees, from 1s. 6d. to 2s. 6d. a lesson. He made analyses, and was consulted by manufacturers, probably the earliest in the district of that class of scientific men called "professional chemists" who have risen as a necessity of the time, and by private establishments have made some compensation for the lack of public institutions and professorships, in some countries so abundant, and have chiefly in their hands the connexion of the chemical arts with the science as it progresses.

Dalton's character as a man is then easy to understand; he was a simple inquirer into nature, his enthusiasm rose only in her presence, his life was devoted to her study. Abstracted in a great measure from the world in its social relations, he was sufficiently connected with it to have endeared himself to all those with whom he lived, and to have formed with some of his contemporaries the warmest friendships. The friends of his childhood were never forgotten, but more warmly remembered as he grew older; whilst he did not the less remember those that he learnt to know only since his manhood. Gentle at least in his spirit, his very solitary life and abstract mode of thinking had not allowed him time to modulate his voice to suit the ears of those accustomed to more polished society, and a certain rigidity of body, as well as of mind, caused the movements of both to have the appearance more of power than of grace. He was simple, temperate, and regular in his habits, never carried away by the feelings of the moment to indulge in luxuries to which he was unaccustomed, in all his actions avoiding excess, yielding to order and regularity as the only master passion which had power to carry him beyond what we may consider the just bounds of reason.

Although unwilling to offend, he was accustomed sometimes to give severe rebukes to ignorance when it pretended to know, and in this capacity alone do we find him ever creating a feeling of opposition in those who surrounded him, although even this was seldom, and only on great provocations, and chiefly in his character as President of the Philosophical Society, where the exercise of wise authority was expected and desired. Local memory tells us of several severe, but well deserved and not ill-natured rebukes.

In a life of labor, of experiments with weights and with numbers, it is seldom that the imagination flourishes, and so we find that literature was entirely neglected, and his own discoveries have not been illuminated by the radiance which it is in the power of some men to shed around the creations of their mind, but have been sent out dry and hard into the world to gather as they best might the life which he was certain of obtaining for them.

In this is the secret of many varying opinions about Dalton; this is the secret of his want of success in early life, of his remaining so long apart from scientific men, and of the disputes as to his originality. The "genial current of the soul" had been constitutionally stopped, and words were wanting to express his feeling, which at last seemed not even to struggle for utterance. He gave us knowledge, mere knowledge does not give life until it is become a familiar inmate of the mind, until we can see it in all or many of its aspects, until it becomes an object on which we delight to gaze, and seeing its beauty begin to surround it with results from the imagination. He that can do this, putting the results of science into a poetic form, and impressing them upon the mind, receives often more admiration than the originator, the exertions of whose whole life may probably be summed up in a sentence, and seen at a glance by the popular eye; but experience proves that the original work requires a persistance of effort and a clearness of conception, which are very rarely united.

True, Davy and Dumas add poetry and eloquence to sterling scientific vigour, whilst in Dalton we find only strength and rude simplicity; the glowing radiance which dazzles us in the writings, and even in the characters of some men, is wanting in him; but strength and simplicity are rare and valuable gifts, although we must look to their combination with beauty in its widest sense for the very highest standard of mankind.

CHAPTER IV.

HISTORY OF THE ATOMIC THEORY.

IDEAS OF MATTER UP TO THE TIME OF LUCRETIUS.

THERE is no chapter in the history of man more marvellous than that which deals with his conception of matter. There has been the greatest difficulty in all ages in comprehending its existence, and still more so in conceiving how it can be constituted of so many different substances. It seems, beyond expression, strange, that although himself made of matter, and exposed so frequently to the pain of living in regions covered with most inhospitable forms of it, or tossed about in an unmanageable ocean of the same, he should still for a long time confuse the conception of it with spiritual existence, and still longer fail to obtain any distinct idea of the cause of their diversity. I do not allude only to the metaphysical difficulties even now unsolved as to the existence of matter, but to those perhaps most apparent in the history of the physical sciences. The difficulties have begun with the savage who scarcely distinguishes the wood from his divinity, and still less divides substances into classes. They are difficulties which, in one form or other, have struggled long in man, and the progress of which may be seen perhaps most clearly in the Greek, but more or less in all nations who have thought on philosophical subjects, whilst the struggle in Europe in the middle ages was long and violent, occupying the most active minds of the age. It is strange to observe the pertinacity of man in deciding that matter is *one*, that all substances have the same substratum without distinct facts as a proof, but relying on his reasoning powers alone; deciding that fire, air, earth, and water, are the same, although burnt out of his house by one, and drowned by the other, and to obtain a third, risking everything that he possesses. He has with difficulty been able to think of the facts before him, except under the influence of some previous conclusion, and pressed to the earth as he has been by physical difficulties, as well as by sensuality, he has continually clothed it with immateriality. Sometimes he appears as a spiritual being, from some higher state of his metempsychosis, with difficulty treating conceptions new to him about mere material things, or perhaps more like a material being oppressed with weakness of conception, grasping at more than he can understand, he has failed to see clearly the facts that of all others seem to stand most prominent before him.

The idea of matter representing the present stratum of knowledge obtained from experiment, and which chemists so long entirely missed, is, that there exist bodies which we cannot divide and call simple, that these by union among themselves form other bodies; that when united the original bodies are by no means lost, and may be again separated without losing any of their original indestructible properties. This statement is a simple relation of facts. These bodies may be farther divisible; they may be, and probably are, all convertible into one, but when we trace them further than our experiments warrant us, we go back to the position of the Greeks, who have already said nearly all that the mind seems able to attain to, unassisted by the study of nature. Indeed, this idea cannot be said to be in any way new, but one of the very oldest, although frequently lost, to be seen at intervals in fragments, but not until within the memory of man to obtain permanent ground in science: although inevitably doomed to prove only a portion of the truth, it is no less true in its own limits. So simple is the idea, requiring too such enormous labour and long time to obtain, that it makes us readily believe that we may now be living under the grossest delusions which a little spark of genius might readily dispel, revealing to us a world entirely different to that which we are accustomed to see; and without a doubt this is to a great extent the case.

As an instance of the difficulty of arriving at a rational conception of material phenomena, but more especially in order to give a sketch of the history of the subject, I shall adduce the opinions shortly expressed of some of the most prominent thinkers of ancient times.

Much as has been written on this ancient part of philosophical history, a full as well as distinct account is still wanting. Most writers give us so much explanation, that we cannot understand them: the collected fragments would form a valuable volume. The physical has generally been given as a mere appendage to the metaphysical history.

Although the earliest opinions relate more directly to creation or cosmogony than to the nature of combinations, they still are interesting in connection with our subject, as they show us the early methods of viewing matter, and illustrate the difficulties as well as progress of the subject. Of the early Greek schools, where fragments only exist, I shall give a few particulars taken chiefly from Ritter and Tiedemann, not giving the original words of the ancient authors, some of which have been lost, and others greatly scattered.*

Thales considered the earth to be a living being, and the only primitive principle from which all things are formed to be water. He observed, no doubt, that where there is no water, there is no growth of vegetable or animal life. The vegetables are fed by the rain; they contain water in their juices, or they do not live. Animals dried up are dead. Life goes on in the world only with water, even the

* Ritter's "*Geschichte der Philosophie.*" I used the French edition. Tiedemann's "Geist der Spekulativen Philosophie."

land grows from it, the islands lift their heads out of water, and the continents repose on great waters.

The rising of the world from oceans was a theme in the still earlier cosmogonies of the east.

If all the solid things spring from water, and if water may pass into air as it seems to do in evaporating, then all things spring from water.

This idea is scarcely dead. Van Helmont believed he proved by experiment that plants grow from water, and without correct analyses, we should be obliged to decide in the same way.

Anaximenes believed the principle of all things to be found in the boundless air. We may reason thus with him, partly in his own words, and partly in ours. The world is limited, the land and water have a definite termination, the air alone is boundless. The air sometimes condenses from itself fierce winds, black clouds, rain, snow, and solid hail; these again are found on the earth as water, and are found to contain the solid world, and although this may rest in the water, the water itself rests in the infinite expanse of air. We breathe air, and live by doing so; when we cease, life ceases; the life and soul are air, which becomes in this way, not only the spirit which moves all things, but the source from which all things are produced, a living principle to the world as a whole, as well as to us.* "He believed in four principle degrees in the qualities of air, which responded to the common opinions of four elements; from these degrees, fire, air, water, and earth, were formed all the other properties of natural things."†

Diogenes, of Apollonia, believed also in "air being the origin of all things, but requires a greater variety of this element. For one is not the same as the other, for there are many varieties of air and many considerations; some is warmer, some colder, some drier, some moister, some calmer, some

* Ritter, Vol. I., p. 182. † Page 185.

more agitated, and it has many other alterations, caused both by its qualities and its substance; and the soul of all living things is air," &c. The air penetrates everywhere, it must be endued with intelligence, as all things are disposed with consummate wisdom. The differences in the sensible qualities of things are referred to condensation and expansion.

To prove that all things are one. "It appears that all that exists is merely the change of one and the same thing; and this is evident, that if all that is in the world, the earth and water, and other things, were different, and made essential changes, there could be no transformations among things."* This is correct reasoning, if we were convinced of transformations, and we must remember that without analysis, we must believe in such changes, as the plant seems to be transformed from water or earth. This reasoning marked a necessary step in the progress of chemistry.

We see here that properties have to be given to air in order to account for its diversities. These properties, among the most important, heat, are not clearly defined. The earth is precipitated from the air by condensation, but there is no explanation of condensation, so that after reasoning on these few principles, the simple atmospheric air which formed the commencement becomes something else, and a mystic air, a kind of substratum of air, is in reality what seems to be signified, as no common air could contain all the properties required. The same is wanted with water, which, after a while, has too much given it to do, to be only common water, and we find the very same thing occur with the rest of the four elements, as well as with the three alchemistic elements.

The soul of all things producing such admirable order, as Diogenes perceived, "it is astonishing that this doctrine did not conduct to the distinction of mind and matter, and that he conceived the principle of all things, even intellectual

* Page 188.

phenomena themselves, to be material." *—Ritter. But however this may be, it is not less astonishing than very much later opinions which distinctly gave material form to ideas. Diogenes was the last of this class who conceived the world to be a living thing, in a narrow sense, and each individual as for a time only able to live isolated, and to resist the influence of the external and greater life which in the end enveloped all.

We have seen air and matter made the first principles, and now we find Heraclitus, of Ephesus, fixing on fire, which before had been made to play an important part as heat, expanding or contracting all things. In the absence of any full exposition of his reasoning, let us rather attempt, as with the others, to complete it for ourselves. He seems to have thought that wherever there is warmth there is life, where heat comes there is motion and activity, it is the principle moving all things, a secret fire which gives life to all things, it produces air, as we see it constantly do when substances burn. Air again produces other elements. As fire produces air, so fire converts water into air, and heat causes those changes in the atmosphere which produce water, whilst by its action on the earth it produces land also, which is raised from the sea. As all things have preceded fire, so in the end must fire swallow up all things or all things will return to it. "The harmony of the world arises from contrary forces, as that of the lyre and the bow."† "The finest harmony is produced by opposites, and every thing is produced by strife." But as with the others, so with Heraclitus, fire was not common flame, but gradually became an expression of force, and we still use the word in this more spiritual sense.

In Anaximander, of Miletus, we find the elements operated on, and transformed by a power which he calls "the infinite," and the higher forces are gradually developed out of the lower. This, although an interesting chapter in the pro-

* Page 190. † Page 214.

gressive theory, which is probably the oldest of all theories, is not sufficiently related to the chemical characters of bodies.

After finding force and intelligence given to matter, we find Pythagoras going to quite the opposite extreme, and making, to all appearance, the origin of all things to be in numbers. Tiedemann says, "In these early days of philosophy, the abstract was not separate from the concrete, there were, it is true, different names for them, but the different meanings were not distinct.* Pythagoras took his abstract and general ideas of objects for the objects themselves, and converted general ideas into substances." Although the Pythagoreans gave much of the work of the world to fire, they had no distinct ideas on the elements.† In harmony with this kind of reasoning, they produced everything from points which seem to have been mathematical, and again, all things were produced from God, because they were produced from numbers, the first of which is "one."‡ This one is the highest God, and of Him our minds are portions.

It is a question whether we can even now in all stages separate the concrete from the abstract. The principle of the Pythagoreans is in reality a pure dynamical theory of matter where there are mathematical points and surfaces forming the limit of bodies. "All things are composed of points or unities of space which form together a number." The use of the word number introduces a difficulty into the conception of the subject, but leads us at once to the conception of force, whatever may be its origin. Their love of music led them to infuse harmony into all nature, and their love of mathematics led them to see in it order and numerical arrangement. They gave the highest place to *one;* they attached particular virtues to the small numbers up to five, apparently for mathematical reasons; they admired seven as the origin of seven chords and planets;

*Tiedemann, Vol. I., p. 96. † Page 97. ‡ Ritter, Vol. I., p. 325.

and ten the number of elementary qualities and their contraries.

On applying these numbers to matter and to calculation, we see no inclination to enter into details and to explain the constitution of different bodies by such means. Or, in other words, the numbers of Pythagoras are not of such a character as to give us the slightest clue either to the atomic theory or that of equivalents, although the remnants of his philosophy shew that he must have greatly exalted the mode of thinking wherever he taught, and led men to seek law and science or great and beautiful truths in the study of nature.

He believed in five elements. This spiritual method of derivation, or explanation, is at least an interesting instance of that mode of thought. Numbers being the origin, the monad, is a point; the dyad (or dual), is a line; the triad, is a surface; the tetrade, a geometrical body; the pentade, the physical body with sensible properties. The cube, was the earth; the pyramid, fire; the octahedron, air; the icosahedron, water; the dodecahedron, the fifth element, Aristotle's ether.

This is enough from the many contradictory notices about Pythagoras, it leads to a pantheistic view of creation, and is another feature in the progress of the subject.* The Pythagoreans spoke chiefly of morals; of nineteen writers, whose fragments I consulted, none spoke of physics.

Anaxagoras says, "The Greeks are wrong in thinking that some things are produced and others perish, for nothing is produced, and nothing perishes; but some things are mixed, some separated, some confused, some distinct, and production or perishing may be properly called composition or mixture, and decomposition or separation."

"The number of things remains always the same." He

* Tiedemann, Vol. I., p. 118.

conceives all the elements to have existed in infinitely small parts. "All things were together, infinite in number and in smallness,* and this smallness was infinite, and all being together, nothing was distinct because of its smallness." All was put in motion by a mind which governed all, and the elements attained differences of character by the preponderance of one or other. The small particles are in constant motion. There are of these particles endless numbers of every character; so a piece of gold consists of endless pieces of gold, so also silver, copper, blood, bones, flesh, water, fire, and earth, consist of infinitely small portions of these substances.† To effect this he adds, that everything was found in everything, because, as the animal grows from its food, the parts of the animal must be contained in its food. A kind of reasoning which, in one sense, is undeniable, but no proper account is taken of the formation of compound bodies, although his ideas of mixed and simple elements might have been supposed to lead him to this mode of reasoning, without the invention of *homoiomereia* (or homeomeria), the name given to this notion of every body being formed of particles like to it.

In this we find a want of discrimination in separating metals from organic substances, the changes of which seem much more allied to transformations, but we find in him philosophy obtaining a view of mind and matter as distinct, and the existence of a ruling power, God, introduced into science. But even here there was some difficulty in obtaining an idea of power without matter, and that is made like an ether, which, however, has an undefined meaning, although it seems generally to refer to a more refined kind of air, expanded into force and intelligence.

It is by no means intended to expound the various philosophies of the ancients, and so from Parmenides we can only

* Tiedemann, p. 251. † Page 1316.

find two ideas exactly suiting the subject. First, all is one. There is only one existence, and there is nothing but existence. Secondly, thought is completeness. Here we may then say that we have got an opinion different from the preceding, in which physical forces have no independent place, although some of the expressions of this philosopher would lead to believe that he considered the earth as the origin of all things, acted on by fire.

Zeno, of Elea, had four elements, warm and cold, moist and dry, corresponding to the four ordinary elements, with necessity as a moving force regulating all; concord and discord (attraction and repulsion) were some of its manifestations; but in reality nothing existed. We have then, one after another, a play on every one of the elements, each elevated in its turn, diminished to one or to a mere idea, or increased to an endless extent, where idea is only the action of an element.

Empedocles gives more distinct form to the four elements, at the same time elevating them by the name of gods. With him they are eternal, and consisting of minute parts, which although divisible, are never divided. This is an early approach to our present chemical theory of a diversity of elements, not transmutable. The principal place is given to fire. But the *four* were upheld logically, when he said they were never divided; but he afterwards adds that they were in reality only two. By him a new phase of character was given to the elements, for he says "our souls consist of all four elements, and every element is itself a soul."* "Life can only be known by life; for by the earth we know the earth, by the water the water, the divine air by the air, the devouring fire by fire, love by love only, and strife by direful strife."† The principle of love (φιλια) he held to be the origin of the elements, and the cause of their unions; the opposite prin-

* Tiedemann, Vol. I., p. 253. † Quoted by Ritter, Vol. I., p. 454.

ciple (νεικος) discord, acting with it, produced the various changes. We have here attraction and repulsion in their early days; but it is also said by him rather curiously, "discord decomposes the mixtures of the elements, and mixes fire with fire, air with air, each sort of element with its like, whilst concord acts on the contraries,"* as now found with electric + and — poles.

Leucippus first distinctly gave a meaning to the notion of small particles of bodies, which he called atoms. Democritus held the same opinions. Everything is composed of indivisible atoms. They could not be divisible, neither could they be mere points. They have neither colour, taste, smell, heat, nor cold: all these properties are given them by their various mixtures: there are various shapes. The first impulse to motion was given probably by an original force.

This is the real meaning of all he said; everything was referred to atoms, even the soul or mind itself. This is a point of great importance in the history of our knowledge of matter, and one beyond which we have not yet got in some of its relations. Here then we stand and are obliged to review our speculations and inquiries in some respect from the standing point of Leucippus and Democritus. In this view we have a distinct idea attached to the composition of bodies, and although one which might have readily come into the mind of any one who thought clearly on the subject, yet we are not aware of the difficulties attending the production of ideas, viewing them after they have been overcome. Democritus arrived at the idea of distinct atoms forming matter of every kind by the change of position. Anaxagoras was nearly at this point, but he gave the atoms characters exactly like the compound object. Democritus gives the simple bodies only shape, extension, and force.

Plato taught that the world was created by an intelligent

* Ritter, Vol. I., p. 445.

cause, and did not exist from eternity, and that it has passed from order to disorder, because order is better than disorder. God also has endowed the world with reason, it is an animated being, and not to be destroyed, but by him who made it; but he will not destroy it, as the good cannot destroy what is beautifully fitted. The world moves by its own life. The elements seem to Plato, to be only forms under which matter exists, and are convertible one into the other. The five forms of matter serve as a base to determine the elementary forms, the pyramid corresponds to fire, the cube to earth, the octahedron to air, the icosahedron to water, whilst the dodecahedron, like the sphere, comprehends like the earth, all the elements.

These are the expressions of Ritter, but on looking into the original it is not found exactly so, although the mode of reasoning somewhat justifies it. The words are "ἔστω δὲ, *κατὰ τον ὀρθὸν λόγον καὶ κατὰ τὸν ἐικότα, τὸ μὲν τῆς πυραμίδος στερεὸν γεγονος ἔιδος, πυρὸς στοιχἔιον κὰι σπερμα,*" &c., as in the translation of Davies. "Let it be agreed then that according both to strict probable reasoning the solid form of pyramid is the *element and germ* of fire," &c.; another way of expressing matter dynamically.

In the Timæus he says, "First, then, that fire and earth, water and air, are bodies, is evident. * * * * We must relate, then, of what kind these most beautiful bodies were that thus came into being, and which, however unlike each other, may yet be produced from each other by dissolution. By accomplishing this we shall ascertain the truth about the generation of earth and fire, as well as those elements (water and air) which hold an intermediate position, for then we shall allow no one to assert that there are visible bodies more beautiful than these, each of which belongs to a separate class." But this does not seem to have been held firmly, because he says also at 51 Timæus, "Our plastic Creator, reflecting on all this, then mingled and united matter, fire, and earth, gradually mixing therewith a ferment of acid and salt (ἐξ ὀξεός καὶ ἁλμυρõυ), and thus he composed a soft pulpy flesh."

As to the composition of the elements, as all was created from mind, the separation of mind and matter does not seem clear. Although he says, also, "It is evident to every one that fire, air, water, and earth are elements, but every species of body possesses solidity, and every solid must necessarily be contained by planes. Again, a base formed of a perfectly plane surface is composed from triangles. But all triangles are originally of two kinds, each of them having one angle, a right angle, and the two others acute, and one of these has an equal part of a right angle, divided by the equal sides, while in the other, two unequal parts of the right angle, are divided by the unequal sides. This, then, we lay down according both to probability and necessity, as the origin and principle of fire and all other bodies; but as for the heavenly principles thereof, those indeed are known only to the Deity, and to those among men who enjoy God's favour."* One might suppose he was speaking of the shape of the elements, and we see the want of definiteness in the forms of conception of the physical elements. Again, matter so far as it is only matter must be viewed abstractedly from all qualities; qualities are determinate conditions of matter. The motions of matter are without rule, purpose, or harmony; purpose, order, and harmony are only to be obtained from the reason.† Physics was in reality the region of uncertainty, whilst true science was in the reason: we have in modern times adopted a different opinion. This will probably be enough to enable us to find the class to which Plato belongs, as regards our subject. By going much further we get into metaphysics, or perhaps worse, into contradictions.

* Timæus, par. 28.

† Tennemann. System der Platonischen Philosophie. Vol. III., pp. 30-32.

In looking on physical nature, as far as our object is concerned, little is got in Aristotle, the idea of which did not exist elsewhere. His first form of substance, which is perceptible to the senses, is finite and perishable; he uses the ordinary four elements. There is a fifth element preceding the four elements, with a tendency neither above nor below; this is ether. The heaven is made of it, and never changes. The four elements seem to be substance united to the warm, the cold, the light, and the heavy.* He more clearly brought forward existing theories, expressing his own with greater care, and giving a history of others.

The varying phases of matter and of force shewed themselves in after philosophies. Matter rose and fell, mind rose and fell. Matter was mind, mind was matter, and even at this period we see no such nice distinction between them among the ancients as we now have, whatever be the foundation of our opinions.

The stoics said, that "matter (that is; considered in itself without quality and form) did not exist except under a certain form, and with certain properties. It is the principle of every thing which springs from it, and consequently is variable. Being absolutely passive it is infinitely divisible as body is."† But these opinions would lead into grounds too little physical.

The stoics retained the four elements which lasted so long in science, and believed that fire was condensed into air, air into water, water into earth. They called matter a collection of dimensions, length, breadth and thickness, leaving out solidity, so that matter became penetrable.‡ Only matter can do any thing. This led to the wildest assertions, that laughing, dancing, walking, crying, as well as emotions, anger, joy, fear and passions, avarice, pride and envy, vices and virtues, day, night, and sound, were bodies. This was carrying

* Tiedemann, Vol. II., p. 284. † Ritter, p. 479. ‡ Tiedemann, p. 434.

out their general principle to the utmost, and, of course, leads to an entire want of definition of natural things. Knowledge is gained by observing qualities, not by confounding them; by seeing distinctions, not by hiding them. The most contradictory assertions of all kinds have been made on the subject, especially after the great masters had done their utmost services. We must not suppose the above to be merely ridiculous. It evidently involves an extension of the idea of body, the limits of which are still unknown, and in some form or other it has often risen, and is likely again to rise, for discussion.

We find that in nearly all the cases alluded to matter has been able to put on various forms, and that it is of itself a mere abstraction. There is, in nearly all, a substratum more or less decidedly expressed. That is, a matter without properties to the senses, but capable of putting on all. The *mater*, mother of all substances. With this idea before us, most of the opinions will have some connection and rationality. Among those who denied the existence of the reality in the things perceived by the senses, we can find very little directly relating to the subject; but we must ever view with admiration and gratitude the acute minds which have done so much of the preliminary work necessary both for physics and philosophy.

The atomic system of the ancients was most fully explained by Lucretius, and leaving the nice distinctions of the stoics, and their semi-metaphysical modes of looking on matter, let us look more fully at this system than the others, as it may be said to form the beginning of the atomic theory, although the short and meagre introduction preceding may not be without interest to such as have not had time to read of the struggles of the mind in early times towards a rational expression of phenomena. It was a struggle of the most gifted minds in some of the most brilliant days of the world, and will never cease to be an interesting chapter in man's history.

The more distinct conceptions of matter introduced by Leucippus, and promulgated by Democritus, were adopted by Epicurus, and have often gone by his name. If not ultimately the most exact, they have in many respects a practical truth, and they have the merit of having the main features clearly intelligible. The system was gaining ground at a time when the Alexandrian school was saying that matter which can be perceived emanates from the soul,* and that bodies were convertible into each other because made of one matter, which original matter had no qualities, and was capable of taking all.† This was very much in the manner of their predecessors, except that we recognise in it an increase of mysticism in their expressions. This substratum of matter has bewildered whole tribes of philosophers.

Lucretius is in the hands of every one, but read by few. The following portion on atoms is from the translation of the Rev. J. S. Watson (Bohn), with little alteration:—

"Nothing can do or suffer without bodily substance, nor, moreover, afford place (*i. e.*, for acting and suffering) except empty and vacant space. No third nature, therefore, (distinct) in itself, besides vacant space and material substance, can possibly be left in the sum of things; no third kind of being, which can at any time affect our senses, or which any one can find out by the exercise of his reason.‡ * * * * * Bodies are partly original elements of things, and partly those which are formed of a combination of those elements. But those which are elements of things no force can break, for they successfully resist all force by solidity of substance; although, perhaps, it seems difficult to believe that anything of so solid a substance can be found in nature; for the lightning of heaven passes through the walls of houses, as also noises and voices pass;

* Ritter, Vol. IV., p. 488.

† Tiedemann, Vol. III., p. 295. See also Histoire de l'ecole d'Alexandrie, par M. Jules Simon. No short sentence can give the exact truth.

‡ Book I., l. 444-449.

iron glows in the fire; rocks often burst with fervent heat; the hardness of gold, losing its firmness, is often dissolved by heat; the icy coldness of brass, overcome by flame, melts; heat and penetrable cold enter into silver, for we have felt both with our hand, when, as we held cups straight in the hand, water was poured into them from above, so that as far as these instances go there is nothing solid in nature. But because, however, right, reason, and the nature of things,. compel (me to hold a different opinion,) grant me your attention until I make it plain in a few verses, that there really exist such bodies as are of a solid and eternal corporeal substance, which bodies we prove to be seeds and primary particles of things, of which the whole generated universe now consists."

"In the first place, since a two-fold nature of two things extremely dissimilar has been found to exist, viz., matter and space, in which everything is done, it must necessarily be that which exists by itself for itself, and pure (free from mixture); for wheresoever there is empty space, which we call a vacuum, there is no matter; and likewise wheresoever matter maintains itself, there by no means exists empty space. Original substances are therefore solid, and without vacuity.

"Furthermore, since in things that are produced there is empty space, solid matter must exist around it; nor can anything be proved by just argument to conceal vacuity, and to contain it within its body, unless we admit that which contains it to be a solid. But that solid can be nothing but a combination of matter, such as may have the power of keeping a vacuity enclosed. Matter, therefore, which consists of solid body may be eternal, while other substances may be dissolved (or cease to be). In addition, too, if there were no space to be vacant and unoccupied all would be solid. On the other hand, unless there were certain bodies to fill up completely the spaces which they occupy, all space which exists must be an empty word. Body, therefore, is evidently distinct from empty space.

"These bodies (which thus fill up empty space) can neither be broken in pieces by being struck with bodies externally, nor again can be decomposed by being penetrated internally; nor can they be made to yield, if attempted, by any other method, which we have demonstrated to you a little above; for neither does it seem possible for anything to be dashed in pieces without a vacuum, nor to be broken, nor to be divided in two, by cutting; nor to admit moisture, nor moreover subtle cold, nor penetrating fire, by which all things are dissolved; and the more anything contains empty space within it, the more it yields when thoroughly tried by these means. If, therefore, the primary atoms are solid and without void, they must of necessity be eternal.

"Again, unless there had been eternal matter, all things before this time would have been utterly reduced to nothing, and whatsoever we behold would be a reproduction from nothing. But since I have shown above that nothing can be produced from nothing, and that that which has been produced can not be resolved into nothing, the primary elements must be a of an imperishable substance, into which every body may be dissolved, so that matter may be supplied for the reproduction of things. The primordial elements therefore are of pure solidity, nor could they otherwise, preserved as they have been for ages, repair things through the infinite space of time.

"Besides, if nature had set no limit to the destruction of things, the particles of matter would by this time have been so reduced, every former age wasting them, that no body compounded of them could, from any certain time, reach full maturity of existence. For we see that anything may be sooner broken to pieces than put together again; for which reason, that which the infinitely long duration of past time had broken into parts, disturbing and dissevering it, could never be repaired in time to come. But now, as is evident, there remains appointed a certain limit to destruction, since we see every thing recruited, and stated portions of time assigned to every thing according to its kind, in which it may be able to attain full vigour of age.

"To this is added, that although the primary particles of matter are perfectly solid, yet that all things which are formed of them, may be rendered soft, as air, water, earth, fire, because there is vacant space intermingled with the things compounded. But, on the other hand, if the primordial elements were soft, how strong flints and iron could be produced, no explanation could be given, for nature would be deprived of all possibility of commencing a foundation. The primordial elements therefore are endowed with pure solidity;

by the dense combination of which all compound bodies may be closely compacted, and exhibit powerful strength.* Moreover, if no limit has been appointed to the dissolution of bodies, there must remain certain bodies in the world which have not yet been assailed with any trial of their strength. But since (dissoluble bodies) are endued with a fragile nature, it is inconsistent to suppose that they could have lasted through an infinite course of time, harassed age after age with innumerable assaults.† * * * *

"Primordial atoms are therefore of pure solidity, which, composed of the smallest points, closely cohere, not combined of a union of any other things, but rather endowed with an eternal, simple existence, from which nature allows nothing to be broken off, or even diminished, reserving them as seeds for her productions.

"Moreover, unless there be some *least*, the smallest bodies will, individually, consist of infinite parts.‡ * * * * What therefore will be the difference between the greatest and smallest of bodies? It will not be possible that there should be any difference; for though the whole entire sum of things be infinite, yet the smallest things which exist will equally consist of infinite parts.‖

"* * * * * Those who think that fire is the original principle of things, and that the universe is maintained from fire alone, do greatly err from true reason; of which Heraclitus, as leader, first comes to the battle, celebrated for the obscurity of his language. * * * * For fools rather delight in all things which they see hid under inversions of words. * * *

"For how, I ask, could things be so various if they were produced from fire alone and pure (from mixture)? Since it would be to no purpose that hot fire should be condensed or rarefied, if the parts of fire retained the same nature which the whole of the fire still has? * * * * But if they think that fire may by any means be extinguished in condensation, and change its natural consistence, and if they shall not hesitate to allow that this may take place absolutely, then all heat, it is evident, will fall utterly to nothing, and whatever things are reproduced, will be made out of nothing. For whatever departs from its own limits, this straightway is the death of that

* l. 484-577. † Not in my copy of original. ‡ l. 603-610.
‖ l. 613-616.

which was before. Something, therefore, must necessarily remain unchanged in that fire of theirs, that all things as you see, may not utterly fall to nothing, and that the multitude of objects in the universe may not have to flourish by being reproduced from nothing.*

† "It would be to no purpose that some (of these elements) should detach themselves and depart, and be assigned to another place, and that some should have their order changed, if they all still retained the nature of fire, for whatever (fire) should produce would be in all forms only fire. But, as I think, it stands thus:—There are certain elementary bodies, whose combinations, movements, order, position, shapes, produce fire, and which, when their order is changed, change their nature; nor, as I think, are they like to fire, or to any other thing, which has the power of emitting particles to our senses, and affecting our touch by its application.

‡ "Wherefore those who have thought that fire is the primary matter of all things, and that the whole universe may originate from fire; and those who have determined that air is the first principle for the production of things; those who have imagined that water can itself form things of itself, and that the earth produces all things, and is changed into all substances of things, appear all to have wandered extremely far from the truth.

"Also those who couple the elements of things, uniting air with fire, and earth with water, and who think that from these four things all bodies proceed. * * * * § Moreover, if all things were produced from these four bodies, and all things dissolved into these bodies, how can these be called the primary elements of things, rather than, on the other hand, things (called the elements) of them, and a backward computation be made?

‖ And now let us examine the *ὁμοιομέρεια* (homeomeria), as the Greeks call it, of Anaxagoras, nor does the poverty of our native tongue allow us to name it in our own language. * * * He thinks that bones are produced from small and minute bones. So likewise flesh is generated from small and minute particles of flesh, and so on.

* l. 636-675 † l. 681. ‡ l. 706. § l. 764. ‖ l. 830.

" * * * * Moreover, since food augments and nourishes the body, we may understand that veins and blood, and bones, and nerves consist of heterogeneous parts. Or if they shall say, that all food is of a mixed substance, and contains in itself small elements of nerves and bones, and also veins and particles of blood, it will follow that both all solid food, and liquid itself, must be thought to consist of such heterogeneous matter, and be mixed up of bones and nerves, and veins and blood. Besides, if whatever bodies grow from the earth are previously in the earth, earth must consist of all these heterogeneous matters which spring from the earth."

He gets however into a similar difficulty, by saying that the atoms which make up these substances, have primary particles of a different figure.

* " Finally, if you think that whatever things you see in the visible world, could not have been formed without supposing the primary particles of matter to be endowed, with a nature similar to the things formed from them, your original elements of things by this hypothesis fall. For the consequences will be that you must have primary particles of matter, which, being the origin of laughter, are themselves convulsed with tremulous fits of laughter, and others which bedew their own faces and cheeks with salt tears."

As to forces, he says:—

† " For certainly neither the primary elements of things disposed themselves severally, in their own order by their own counsel or sagacious understanding; nor assuredly did they agree among themselves, what motions each should produce; but because being many, and changed in many ways, they are for an infinite (space of time) agitated, being acted upon by forces, throughout the whole, they thus by experiencing movements, and combinations of every kind, at length settle into such positions, by which means, this sum of things produced, exists."

He speaks again of their being moved of themselves, and urged by secret impulse; and gives their original motion to be a falling straight down, not in a right line. For if they

* l. 914. † l. 1020.

had fallen in a right line, there "would have been no contact produced, and no collision generated among the primary elements."

He gives a limited number of shapes to the atoms, but sees no need of hooks to keep them together. This variation in shape leads also to a variation in size, but the atoms themselves are infinitely numerous.

Primordial atoms are not sentient, or they would "produce nothing but a crowd and multitude of animals."*

So far Lucretius. This theory of the constitution of material substances requires us only to conceive of one class of substances: for although some are larger than others and differently shaped, that is not a necessary supposition. It is the atomic theory properly so called. Forces are as much left out as possible.

One of the most complete atomic systems seems to have been produced in Hindostan. There it is said matter consists of the smallest possible bodies which are indivisible. We must at last arrive at something limited, otherwise the smallest, as well as the greatest, would be infinite. The first compound is binary, the union of two being the simplest, then there is a formation of three binary atoms, and a new compound of four quaternary atoms, and so on. The atom is equal in size to the sixth part of a particle seen by the sun's rays. A superior force draws the atoms to each other. The union is not a mere juxta-position, but one drawn by a particular affinity.

This is a peculiar mode of combining the atoms, we have not three or four simple as we might expect. It is interesting to find exactly the same course of reasoning about natural things among the Hindoos as in Greece, although Mill, in his history of India, ridicules it as the wanderings of the mind; the same might be said of the Greek ideas. We find there,

* Book II., l. 917.

that "from intellect arose fire; from ether, air; from air, fire and light; from light, a change being effected, comes water with the quality of taste; and from water is deposited earth with the quality of smell." To fire is attributed the quality of figure: Mr. Mill supposes light is meant; to air the quality of touch; to ether the quality of conveying sound. Mr. Mill thinks *hearing* is meant. The qualities are confused as in that Greek system, which said, fire only can understand fire; air only can understand air; and so on.* The Hindoos, also, had their four and their five elements; their eternal elements; and the elements proceeding from the will of God, to cease when he pleases. They, too, had a system which made mind a substance and the affections subtle bodies.† But they did not always confuse it, as is seen in Menu. "He having willed to produce various beings from his own divine substance, first with a thought created the world." On the constitution of matter we see them speaking as plainly as the Greeks; and now on the philosophy of the combination, we see an instance of still greater farsight. We can readily believe that the following is very beautiful in the original poetry. There is a poetical beauty even in the prose, which makes it dance like the power it describes in spite of the gravity and intellectual nature of the subject.

From the poem of *Shi'ri'n* and *Ferha'd;* or *the Divine Spirit and a Human Soul Disinterestedly Pious.*‡ "There is a strong propensity which dances through every atom, and attracts the minutest particle to some peculiar object; search this universe from its base to its summit, from fire to air, from water to earth, from all below the moon to all above the celestial spheres, and thou wilt not find a corpuscle destitute of that natural attractability; the very point of the first

* History of British India. By James Mill, Esq., 1848. Vol II., pp. 94-95.

† Colebrooke, Asiatic Researches. Vol. IX., &c.

‡ The Works of Sir William Jones, 1799. Vol. I., pp. 170-171.

thread, in this apparently entangled skein, is no other than such a principle of attraction, and all principles besides are void of a real basis; from such a propensity arises every motion perceived in heavenly or in terrestrial bodies; it is a disposition to be attracted, which taught hard-steel to rush from its place and rivet itself on the magnet; it is the same disposition, which impels the light straw to attach itself to the amber; it is the quality, which gives every substance in nature a tendency toward another, and an inclination forcibly directed to a determinate point."

CHAPTER V.

FROM LUCRETIUS TILL THE DECAY OF ALCHEMY.

EGYPT seems to have attained to a knowledge of the qualities of matter never attained in the Greek and Roman empire, and it is highly probable that a more correct theory existed in conformity with the more advanced practice. But it concealed its knowledge of science as it concealed its history, and the treatises which profess an Egyptian origin are more unintelligible than hieroglyphics. One is at first inclined to believe that the mystic mode of writing was simply a result of ignorance, because when the Alexandrian school begins to philosophize, it uses language as clear as it is capable of obtaining, although prevented from great clearness by the mystic nature of its region of thought. But the results of the arts exist to tell us that in the region of actual knowledge they were no triflers, whilst in that of speculation they made such efforts to gain a knowledge of things superhuman, that if they have failed it is not for want of devotion, truth, and energy.

We have seen matter viewed in various aspects according to the philosophies of the time, and we might almost have added according to the speculations, viz., those of the popular beliefs and the mythologies of the time, when substance was strangely mingled with spirit, and when the gods of the woods were scarcely separated from the woods themselves. Such a view is analogous to the ancient Scandinavian method of looking at creation, when they formed the earth from the body of Ymir, and the sea from his blood.* It shews us that the intellect does not readily conceive of mind and matter, of

* Prose Edda: Mallet's Northern Antiquities.

force and substance, as distinct; that in fact the early mind looks on matter in an abstract sense as a thing made of qualities; which qualities may change indefinitely, and all things easily be transformable. The elements in this philosophy are not elements of science, but the common elements of every day life; in other words those aggregate objects with which nature has made the world, without reference to their mechanical or chemical disunion. In reading the ancient authors, one is disposed to think that this was the general sense in nearly every system, and that a strict meaning was not attained by any one of them, except the atomists. In vain do we quote their opinions, another quotation comes with a meaning in exact contradiction, and the vague and indefinite is the only final result. It is not to be wondered at that this should occur with the Alexandrian school, which neglected the body, despised the world, and sought truth only in that state of mind called ecstasy, which, however exalted it may be, is very naturally shunned by us as a dangerous forerunner of the loss of reason, or as a state of hallucination. With them, there are four elements, it is true, but these are like the elements of Plato and others, they are convertible, and there is an abstract matter from which all things are made by the putting on of various properties. It is the origin of all things that exist, and has the power of becoming everything, but it exists only as a power. This leads into metaphysics, which I avoid; for us, it is enough that they looked on the four elements as transformable, which have, therefore, not the character of elements. What are they, then? They are nothing real, the qualities are changeable, but there are no data given for finding whence these qualities come, and so we are led into a region of mere dreams, and nature is a power playing upon us every imaginable illusion by means of its forces; "all which forces have their origin in *one*, because unity is the basis of all things." Here, then, being no beginning of matter, properly so called, it is in vain to look for a beginning of such elements as cannot be convertible, even if they had not expressly stated that all are capable of transformation.

The early stages of chemistry and metaphysics have unconsciously met. From Alexandria, with its spiritual or mystic views, the *ars sacra*, ἱερα τεχνη, came, perhaps carried by some who had little to teach, at least speaking of matter in terms too mystical to bear to us any distinct meaning, and holding their knowledge too sacred to be given to ordinary mortals.

The spiritual faculty was becoming developed in man. It had existed before that time in its greatest powers in individuals, but it had not as yet educated nations. Christianity had directed man to the consideration of God and the moral part of our nature with such power and success, that philosophy was carried away by the current, and at length became entirely absorbed. But to philosophize on morals and upon spiritual phenomena, so to speak, is to become mystic, unless great care be used; and this current of thought induced, acted a long and important part in the history of the world. The course of this philosophy, for many ages, seldom led it to touch on matter with any firmness of step. Man lived in his thoughts, still fancying himself capable of living independent of mere external things, his spirit still disinclined to believe in the great power which the base creations under his feet were capable of exercising over his finest feelings. There was a struggle against matter, one which cost thousands of victims, buried or slain in the ranks of asceticism, or made useless by the perverted consciences which they carried with them into common life. Matter was not seriously believed in, it was a strange power, a thing capable of every action, acted on by the Deity, or by the spirits of those who either in this world, or the next, had obtained a share of the primeval influence of the Creator. With such men there could be no true philosophy of matter. This mystic mode of philosophizing had no doubt been the mode adopted by the chemists of the earliest centuries, but little or nothing now remains. The following is from Zosimus, the Panopolitan, upon the "Divine water:"—

"The mystery sought is great and divine, for all is from it, and by it. There are two natures and one substance. The one carries off and subdues the other. This is silver-water (mercury), the male-female principle always escaping, constant in its properties, the divine water unknown to the world, the nature of which is inexplicable. For it is not a metal, nor is it water; it is always in movement, nor a body; it is all in all; it has life and spirit, and may be held." *

This is quite in the mystic style of the neoplatonic philosophy, although the tone is evidently lower than that taken by the philosophical mystics themselves. We see in it clearly the style of the alchemists, and the impulse that was given to the study of alchemy seems to have come from this source mainly. The philosophy of Aristotle having for many ages taken the lead, the four elements followed him, and we find them acting an important part in nearly all alchemical treatises, but the mode of reasoning is by no means Aristotelian, and is in every sense mystic. And why mystic? This word is itself of indefinite meaning, but I would say that the reasoning on physical law is mystic, when there is no distinction made between the laws of material nature and those of the mind of man, when there is a confusion between that which is done by natural law and that which is caused by the spirit of man, during his observation of the phenomena; when, for example, an experimenter must suit his frame of mind to the experiment, or find that nature will refuse to act. It may be said that such a theory involves the continual presence of God, supporting constantly his own laws, as in a Hindoo system, † or it is a theism which makes God everywhere acting, not

* Hoefer. Historie de la Chimie, Vol. i., p. 259.

† Blakey's History of Moral Science.

according to fixed laws, but according to man's deservings. The latter was not much carried out, and the first corresponds more closely to the general opinions of these writers. It is a kind of physical mysticism also to confuse elements with each other, and powers with substances, a state corresponding exactly with the mysticism of a more mental character, which confounds the Deity with his creatures, and desires to absorb embodied man in his unembodied creator.*

The four elements were preserved amongst the Arabians, and their corresponding qualities, hot, cold, dry, and moist, and the most wonderful power was given to a mixture of them all; it was in this way that "*Hai Ebn Yokdan* was produced without father and mother; it chanced that a certain mass of earth was so fermented in some period of years, that the four qualities, viz., *hot, cold, dry, moist*, were so equally mixed, that none of them prevailed over the other."† This was Avicenna's opinion of what was possible.

Fermentation was a favourite method of explaining difficult phenomena, and is now. As it has now been to a great extent explained, the lovers of the occult will be obliged to seek deeper for their theories.

We may find then another origin for the mystic mode of viewing matter, as well as for the four elements, as far east as Hindostan, from which the Arabians may have brought it; but we cannot, with historical certainty, trace it to its home; and it is possible that these two origins, Asiatic and Egyptian, may have been originally the same. But what is clear to me is, that the style of the alchemist had the same origin as the

* Whilst this Memoir was being printed, I obtained the "Hours with the Mystics," by Robert Alfred Vaughan, B.A., who says, vol. 1, p. 26; "mysticism, whether in religion or philosophy, is that form of error which mistakes for a divine manifestation the operations of a merely human faculty." This has no doubt influenced the expressions used above. Hoefer has observed the connection between the religious writers of early centuries, but has not traced it up.

† The Improvement of Human Reason, exhibited in the Life of *Hai Ebn Yokdan, by Abu Jaafar Ebn Tophail.* Translated by Simon Ockley, A.M.

mystic style introduced into religion and philosophy. The same condition of mind would produce both, according to the subject to which it was applied, and that condition seems to have a historical as well as natural connection.

Albertus Magnus followed closely the reasoning of Avicenna. He was born in 1193, a few years earlier than Roger Bacon. Some of his opinions on matter shew distinctly their origin. He says,* "That matter and power are the principles of each body is clear from the reasoning; for having taken away all the accidental forms, we arrive at length at a substantial form, which being removed by the intellect, there remains a something very occult, which is the first matter." (Tandem venitur ad formam substantialem qua adhuc abstracta per intellectum; remanet quoddam valde occultum quod est prima materia.) He says also, cap. iii., "Matter has a natural appetite for form." He discusses whether the metals are generated, and decides by a peculiar reasoning that the *materia prima* is not generated, but created, "because if it were generated it would be from some other matter, therefore matter would contain other matter, and so on without end; therefore the *first matter* is not generated, but created. Creation means to make out of nothing." But the elements may be generated, and Albertus Magnus easily changes one into the other. "The generation of one is the corruption of the other, and *e converso*. From the generation of one then follows the corruption of the other."†

That the four elements held their ground we see from no one more clearly than Roger Bacon, born 1214. We see there, too, the opinion that the original matter, *yle*, has none of the qualities of a body, but is matter in the abstract. His words are:—

"Elementa sunt quatuor, ignis, aqua, aer, terra, modi id est

* In the portion of his work on natural philosophy, entitled "Physicorum," cap. 2.

† His chapter, *De Generatione Elementorum*.

proprietates, sunt quatuor, calor, frigiditas, siccitas et humiditas; et yle est res in qua non est calor, nec frigiditas, nec siccitas, nec humiditas et non est corpus. Et Elementa sunt facta de yle; et unumquodque elementorum convertitur in naturam alterius elementi et omnis res in quamlibet. Nam hordeum est equus per vim, id est, naturam occultam; et triticum est homo per vim, et homo est triticum per vim."*

"There are four elements, fire, water, air, and earth, that is the properties of their condition are four, heat, cold, dryness, and wetness, and yle (the true matter) contains no heat, nor cold, nor dryness, nor wetness. The elements are made of yle, and each of the elements is converted into the nature of the other element, and everything into anything else. For barley is a horse by possibility, that is, occult nature, and wheat is a possible man, and man is possible wheat."†

This explanation is exceedingly clear and rational, and founded on a good deal of observation. We see here in Roger Bacon's ideas, which are truly in the spirit of the ancient philosophers, the atomic being excepted, that all being might arise from this source, called *yle*. There is, however, an improvement in the mode of expression; we understand perfectly what he means, and there is a terseness seldom, if ever attained, either by those before him, or after him. It separates him from the mystics, properly so called, although there is the easy passage of one element into the other, and the unsubstantial principle from which all are made.

To this scholastic form the mystic method attached itself, coming apparently with the Arabian learning into Europe, as well as by direct transmission through Greece and Rome, having lived for some centuries with little growth, and destined to be led with more vigor towards its extremes by the reviving intellect of the West.

* De Arte Chymiae. † This yle is the Greek ὑλη, matter.

The method of viewing matter being sufficiently confused by any one of the systems, was not less so when they were united, or rather, I may say, carelessly mixed and confounded. When we add to them an infusion not only of the moral and religious method, but of morals and religion themselves, as elements in the production of results, we have a system of nature which all minds refuse to contemplate with patience, and on which our imagination refuses long to rest, so bewildering is it to the one, and so wanting in beauty to the other. But in a history of man it will always form an important chapter; as a dissection of the mind it will always have a psychological value, and as a portion of the progress of physics it will never cease to be worth preserving for at least one lesson to the student.

The Arabian chemists who took the lead in introducing it, acknowledge Geber to be their master, and he composes metals of sulphur and mercury. At the same time he says, "We see no ox transformed into a goat, nor any one species transmuted into another, or by any other artifice so reduced. Therefore, seeing metals differ in themselves, can you transform one into another, according to its species, or of such a species make such a species? This seems to us sufficiently absurd, and remote from the verity of natural principles. For nature perfects metals in a thousand years; but how can you, in your artifice of transmutation, live a thousand years, seeing you are scarcely able to extend your life to a hundred?" He considers the work as done by the stars, which cause the generation and corruption, but being ignorant of their power, we cannot use it. "Likewise, also in things natural, this is the order; it is easier to destroy them than to make them. But we can scarcely destroy gold, how then can we presume to fabricate the same?"

He was, therefore, not a gold maker, although holding the abstract theory of the possibility of transmuting. His followers did not follow him in this, and whilst quoting his authority, probably knew nothing of his more rational practice; they evidently retrograded, but it was probably needful, in order to satisfy themselves.

A few specimens of the Stoechiometry of the period onward to Boyle will be needful to explain to what a woeful state the science fell; or shall we say rather what a dreadful struggle it had to come to the light; or shall we say that it was only in analogy with the general opinions of the times, and a picture of the general state of mind? For the honor of human nature I would prefer to say that the highest minds being engaged in the cultivation of the more spiritual faculties, in the development of the moral nature of man, in struggling for freedom of body and of mind; the lower ones were engaged with science, and lived a miserable existence, fed by the crusts from the table of the brighter intellects. We may add also that science was, with many persons, as now, a mode of making money only; and many who had no love for it, joined in the pursuit, and are now by us apt to be confounded with true men. This explains many, but not all the cases.

Science and religion have always influenced each other, and it is interesting to trace them. In the *Tractatus aureus*, by a German *philosopher*, the piety is seen. In making experiments, he says to *every pious God-fearing chemist:* "Above all it is needful to be pious, to lift the heart to Him, with true, ardent, and not doubtful prayer, and to ask the gift from Him only."* Again, of the *Quinta essentia.* "It is the universal and scintillating fire of the light of nature, which has the celestial spirit within it animated in the beginning by God, and penetrating all things, called therefore, by Avicenna, the spirit of the world. For as the soul is found in all the members of the human body, and moves itself, so this spirit is found in all the elementary

* Museum Hermeticum, page 80.

creations, which is also the indissoluble connection of body and soul, and so is a most pure and noble essence, full of wonderful efficacy and virtue, in which all mysteries lie hid." *

What this was we see better in the next quotation from Mehung. "Now I consider you informed of this valuable truth, that there is no element which is not worked into another, so that operating on one, the other is operated on. For example, fire is worked into air and earth, if fire excites the operation. The earth is the mother and sustainer of all things; as all things under heaven which are subject to putrefaction, are produced by birth in the warmth of its womb. Only the power of God allows me to return the four elements again into the fifth essence, and that is called the *prima materia*, which is mixed generically in every element." † (Quae in uno quolibet elemento generice mixta est.)

The idea that all things grew was a common one to them, as well as to Dr. Johnson, of modern times, and of great learning. That every thing grew from seeds, is said to be the most ancient of doctrines.‡ There is no detail as to whence the increase of volume came; growing was the cause and the explanation. As it is said in "*The way of Truth,*" "so long as you boil, so long you putrefy it, and the substance is exposed to putrefaction, like corn which is thrown into the earth, and which is preserved in the earth itself by the heat of the sun, but must putrefy by natural rain before anything new will grow from it." ‖

And Basil Valentine, "on the philosopher's stone," says, "the vivifying power of the earth produces all things which spring from it, and he who says that the earth is without life, speaks contrary to truth. For the dead can supply nothing to the living, and neither can the dead grow, because the

* Museum Hermeticum, p. 84.

† Id., p. 151. Demonstratio Naturae. By Mehung.

‡ Aristotle, quoted by Ritter, Vol. i., p. 180.

‖ Idem p. 186.

spirit of life has fled; therefore the spirit is the life and soul of the earth, it dwells in it, and is operated on from celestial and sidereal, into terrestrial; for all herbs, trees, and roots, as well as metals and minerals, receive their strength, increase, and nourishment from the spirit of the earth." * Although this leads to conclusions beyond the known facts, it is a style of reasoning not to be entirely found fault with, and much less can it be called irrational. It is a kind of middle path between calling earth an animal and our present opinions; but Basil Valentine showed intellectual vigor, and formed an epoch in his science. He was born about 1413.

Raymund Lully, who takes us back again to the age of Albertus Magnus and Roger Bacon, was universally quoted by alchemists as a great master, he was born about 1235. He writes;

"Matter says I am a being from which something is made by *passionhood*,† and that substantially and accidentally, because I am the leader, since from me who am the primitive is made that particular matter which is the substantial part of substance, as the matter of a rose, of a horse, and so on.—God is my end and cause, and I am simply his effect.—

"Matter says, I am absolute passionhood under absolute form to which I am conjoined, and as all the river waters are derived from the sea and return to it, so from me are derived all matters in particular because I am absolute."

"Again matter says, I am not a being absolutely existing, potentially so; because, if so, the subject in which I am would be sustained potentially, and so successively *ad infinitum*, which is impossible. I am, therefore, a being existing potentially to all particular substances existing under particular forms."

"In my nature there is not found form, which is from me

* Museum Hermeticum, p. 403.

† I imagine this word to imply better than any I know the state of ready excitability to impressions.

and on account of me, because if so I should not be absolute *passionhood* and my parts would be deprived by me of essence and nature, which is impossible since I am simply absolute."

"Matter says, I am individualised by quantity, as when I am so much; either long, broad, or deep; either round, in a circle, or spherical body; or qualified, for example, when shining or warmed in flame and reddened in red wine, sweetened in honey, heavy in earth, or become light in the fire."

Some may prefer the original of this:—

Ait materia: ego sum ens ex quo fit aliquid passionando et hoc substantialiter et accidentaliter, quià dux sum, quoniam ex me quae sum primitiva fit materia particularis quae est substantialis pars substantiae. * * * * *

Ipse (Deus) est meus finis, mea causa prima, et ego sum effectus simpliciter ejus.

Ait materia, sum absoluta passio sub absoluta forma sub qua sum conjuncta et sicut ex mari derivantur omnes aquae fluviales et ad istud revertuntur, sic a me derivantur omnes materiae particulares et ad me revertuntur, quià absoluta sum. Rursus ait materia. Non sum ens existens in potentia absolutè, quià si sic subjectum in quo essem, sustentata esset in potentia et sic successivé in infinitum quod est impossibile. Sum ergo ens existens in potentia ad omnes substantias particulares sub formis particularibus existentes.

In mea matura non invenitur forma quae sit ex me atque propter me, quià si sic, non essem passio absoluta et secundum quid meae partes essent à me ab essentia et natura privatae, quod est impossibile, cum sim simpliciter absoluta.

Ait materia. Individuata sum per quantitatem cum quae sum quanta, ut puta longa, lata, et profunda, aut in circulo vel in corpore sphaerico rotunda: sum qualificata, ut puta in flamma lucefacta, calefacta; et in vino rubro rubefacta, in melle dulcificata, et in terra ponderosa; et in igne levificata.*

He recognises the four elements, and believes them to exist

* Raymundi Lullii Opera—Philosophiae Principia—De Materia. Strasburg, 1609. I suppose more than one mistake to occur in the Latin. I may be wrong, but it is quoted correctly.

in all things *elemented.* We see in him great acuteness of thought and much valuable observation, and need not wonder that he was a great master among succeeding alchemists. We see in him the Greek mode of reasoning on the abstract principle of matter, but the word which was intended only for a general name, comprehending all substances, is taken for an existence by itself. I know that there are other means of getting into the same difficulty, but we find the alchemists haunted by the ghost of the fine abstractions of Plato's divine ideas of things existing on earth, and the belief in abstract matter, and not seeing the intellectual origin they seek to find them out in substance, and so we have a search for the *prima materia*, which orginally had no power of being handled, and scarcely of being conceived.

They tried in vain to hunt the ghost down, and to handle it in their fingers. Whilst the intellect was unable to grasp the idea, they actually sought to catch it in a bottle. This misconception, so singularly ludicrous, makes me suppose that the class of minds generally engaged in chemistry were inferior, but it may be enough to suppose that the mistake once made was not easily rectified. It seems to me to be without doubt that this is the origin of the singular alchemistic chase. If "immaterial substance" was a "philosophical imposture in philosophy," as Coward called it last century, how much more has it been to alchemy. Yet, I know a living and *intelligent* man who searches for the *prima materia.* It may truly be said that the world is governed by ideas.

This *matter* having become on one side a mere substance, it is not to be wondered at that on the other side it became entirely the opposite, and was a representative of abstract force, as we may see in expressions of Plotinus, and others of the time, which it would take too much time to quote.

Again, we have a search for the fifth essence, which contained all things and was made up of all things. Here is an evident proof of a decrease both of intellectual power and of knowledge, and an illustration of the danger of a little learning.

This *prima materia* was even supposed by some to be the substance on which it was needful to begin to operate, so far had it descended. Flamel says, "Having thus obtained this delicate and precious book, I did nothing else day and night but study upon it, conceiving very well all the operations it pointed forth, but wholly ignorant of the *prima materia* with which I should begin, which made me sad and discontented." * Although this, as is afterwards shewn, was a matter of preparation also, the point of departure of the second process.

Artephius, who lived in the twelfth century, and was born, according to his own account, in the second, and wrote a book on prolonging life at the age of 1052! may be quoted as a true mystic chemist, shewing us, too, by the numerous names that he gives to this liquid, that he even at that time inherited it from a long ancestry. "Our dissolving matter, therefore, carries with it a great tincture, and a great melting or dissolving; because that when it feels the vulgar fire, if there be in it the pure or fine bodies of sol or luna, it immediately melts them, and converts them into its white substance, such as itself is, and gives to the body colour, weight, and tincture. In it also is a power of liquefying or melting all things that can be melted or dissolved; it is a water ponderous, viscous, precious, and worthy to be esteemed, resolving all crude bodies, into their *prima materia*, or first matter, viz., into earth and a viscous powder, that is into sulphur and mercury." The indefinite notion of the origin of properties we see, when he says, "the property, therefore, of our water is, that it melts or dissolves gold and silver, and increases their native tincture or colour; for it changes their bodies

* The Lives of the adepts in Alchemystical Philosophy, with a Critical Catalogue of the Books in this Science, and a Selection of the most celebrated Treatises on the Theory and Practice of the Hermetic Art. Page 34.

from being corporeal, into a spirituality, and it is this water which turns the bodies or corporeal substance into a white vapour, which is a soul that is whiteness itself, subtle, hot, and full of fire."

Here is an attempt to separate chemical action from mechanical. "It appears then that this composition is not a work of the hands, but a change of the natures; because nature dissolves and joins itself, sublimes and lifts itself up." The nature of the treatises which he studied is seen in a quotation of his own. "It is also the most acrid vinegar, concerning which an ancient philosopher has said, I besought the Lord and he shewed me a pure clear water, which I knew to be the pure vinegar, altering, penetrating, and digesting. I say a penetrating vinegar, and the moving instrument for putrefying, resolving and reducing gold or silver into their *prima materia* or first matter." *

That the chief aim was this originally metaphysical idea of matter, we see also as late as 1691, in the Second Aphorism of "Baron Urbiger," whose more modern style has helped him to a more condensed expression. "An indeterminate matter being the beginning of all metals and minerals, it follows, that as soon as any one shall be so happy as to know and conceive it, he shall easily comprehend also their natures, qualities, and properties." "It is found every where, at all times, and only by our science."

A favourite mode of representing chemical action, was by the analogies in the growing of plants and animals. Zosimus says it is the (Hydrargyrum) water-silver, the male-female principle, the principle always escaping constant in its properties. The same idea proceeded forward, and in 1409, Nicholas Flamel says, "minerals taken out of the earth may be changed if beforehand they be spiritualized, and reduced into their sulphureous and argent-vive nature, which are the

* Idem. The Secret Book of Artephius.

two sperms, composed of the elements, the one masculine, the other feminine. The male sulphur is nothing but fire and air; and the true sulphur is as a fire, but not the vulgar, which contains no metallic substance. The feminine sperm is argent-vive, which is nothing but earth and water," &c.

And Norton, about the same time, reasons thus;

"Metalls of kinde grow lowe under grounde,
For above erth rust in them is found;
Soe above erth appeareth corruption
Of metalls, and in long tyme destruction,
Whereof noe cause is found in this case,
But that above erth thei be not in their place." *

But he denies, like Geber, any growing in glass vessels;

"For cause efficient of mettalls find ye shall,
Only to be the vertue minerall,
Which in everie erth is not found,
But in certain places of eligible ground."

When once grown, however, they could not multiply according to him;

"Trewly ye maie trust as I said before,
How of one ounce of silver, maie silver be no more."

But this was not the general opinion, as they were supposed to grow in the preparation, as Ripley says; On calcination, v. 15; †

"If thou intend therefore to make
Gold and sylver by craft of our philosophy;
Thereto nother eggs nor blood thou take,
But gold and sylver which naturally,
Calcyned wysely, and not manually,
And new generation wyll forth bryng,
Incresying theyr kynde as doth each thyng."

As metals grew by their life, and life is the blood, so it was a desirable thing to find out the blood of metals;

"The true blood of mettalls is hard to have." ‡

As the *prima materia* was of indeterminate properties, was

* Ashmole's "Theatrum Chemicum," 1652, pp. 18, 19, and 20.

† Same, p. 132.

‡ "Anon." Same, p. 406.

in fact originally only an intellectual conception, come to in the consideration of matter, abstracted from its apparently unnecessary properties, and converted into a reality by the alchemists, so they, in their turn, converted their elements, which are partly the usual four, partly sulphur and mercury, or with salt added, into mere abstract ideas. Their sulphur was an ideal sulphur, and Paracelsus gave every body its own peculiar sulphur and other elements. This was a step towards a recognition of many distinct indestructible bodies, and a progress out of the purely mystical consideration of the subject.

The opinions held generally by alchemists, having arisen, as I think, from the spiritual and religious state of man, more than his directly intellectual, it would require more space than can be given here for a history such as would suit this view of the case. The early Greek manuscripts, mentioned by Olaus Borrichius, and partly given by Hoefer, show the inclination there was to write with, or to proceed from, the current religious opinions. Isis and Osiris were important alchemistic names from the Alexandrian school, and salt, sulphur, and mercury, have been connected with the trinity of Christians.

It is not the place here to follow the detailed opinions of the alchemists, they are taken rather in the mass, and twelve or fourteen hundred years of their opinions thrown with little order together. This I see no great reason to alter, the men differed so little in their radical opinions. Salt, sulphur, and mercury, were sometimes the origin of the metals, sometimes of all things, each had its own spiritualization, each was the greatest or the least in its turn; and as a curious instance of recurrence of opinion, we find Palissy, in the 16th century, a man full of shrewdness of observation, saying, as Thales did so long before, "that the commencement and origin of all natural things is water." * But then he believes in more

* p. 217. Oeuvres Complètes de Bernard Palissy. Edition par Paul Antoine Cap, 1844.

than one kind of water, one of which he considers a fifth element. This, however, has to do with the history of the idea of liquid and fluid, which must not be entered on here. In saying this, Palissy was reasoning from the state of metals in solution, for he believes they grew, and were found in common water. As he says;

"Il (le createur) a commandé à nature de trauailler, produire et engendrer, consommer et dissiper; comme tu vois que le feu consomme plusieurs choses, aussi il nourrit et soustient plusieurs choses; les eaux debordees dissipent et gastent plusieurs choses, et toutefois sans elles nulle chose ne pourroit dire ie suis. Et tout ainsi que l'eau et le feu dissipent d'vne part, ìls engendrent et produìsent d' autre. Suyuant quoy ie ne puis dire autre chose des metaux, sinon que la matiere d' iceux est vn sel dissoult et liquifié parmy les eaux communes, lequel sel est inconneu aux hommes; d'autant qu' iceluy estant entremeslé parmi les eaux, estant de la mesme couleur que les eaux liquides et diafanes on transparentes, ìl est indistinguible et ìnconnu a tous; n' ayant aucun signe apparent, par lequel les hommes le puissent distinguer d' auec les eaux communes."*

Matter then appeared to the alchemists as it has done to the greater part of early thinkers, and to most of those who do not think, as a power subtle and changeable, capable of every transformation, and dependent on laws by some held eternal and immutable, by others dependent on the spiritual condition of man. We have no distinct clue in such opinions to any definite ideas about composition.

* Same, p. 194. In his objection to alchemy, Palissy has not shone so much as in his pottery, and even in his other scientific inquiries. Bötticher, also a potter, a less honourable and less able man, has been, by his discoveries in Dresden porcelain, the occasion of a saying suitable to the times, and shewing the true value of the alchemists, viz. ; their accidental discoveries in the arts. When his gold making failed, and he made porcelain, it was said—

O Gott Du grosser Schoepfer,
Aus einem goldmacher wird ein Toepfer.

Ye heavens, alchemy has win my votes,
A goldmaker's changed to a maker of pots.

p. 303. Hist. Crit. Untersuchung der Alchemie. By Wiegleb, 1791.

On the four elements we have occasionally opinions which may be called definite, as when they are said to form various combinations with various properties, but these are held with no tenacity. In the salt, sulphur, and mercury, we have a clue to definite composition, but we have seen them spiritualized and held with still less tenacity. In the elements of Leucippus, Democritus, and Lucretius, we have a definite view taken of the subject, but that was soon neglected, to be revived in part when alchemy was falling. We have had an approach to a distinct view before us in that of forces, or unextended points, forming the origin of matter, brought forward later, and, perhaps, in an entirely original form, but at least more detailed, by Boscovich.

CHAPTER VI.

OPINIONS DURING THE TRANSITION FROM ALCHEMY TO CHEMISTRY.

THE ages so rapidly passed over comprehend two thousand years. The attainment in chemical science has been as yet small. All the theories have been abstract; they have been efforts of the mind to comprehend matter, with a very meagre, if any, classification of phenomena. We might ask ourselves if this was caused by the scarceness of facts, or by the want of a mind to perceive them. As to scarceness of information this may be doubted, numerous truly chemical arts had been from early ages in use; many had fallen into disuse. The observers that now come to the science are not so much characterized by the multitude of their discoveries, as by the minuteness of their observations, and the penetrating nature of their reasonings. The progress of mankind has often been compared with the progress of individuals, and the analogy serves here to point out a cause. In the history of every mind, especially of the mind of the student, there are seasons where facts are collected and books read with diligence, theories formed in great numbers, reasonings adopted and thrown aside; opinions are heaped up, but no distinct opinion is formed, and a vague stare into the difficulties before it is the only result of the labour, for knowledge it can scarcely be called, where nothing is well arranged, and nothing is actually known. The imagination then takes the place of the helpless reason. A poetical mind may find in this state a field for its highest powers, a weaker mind will seek explanations more or less mystical, and a still weaker mind will be contented with the merely mysterious. But a vigorous mind, in which reasoning and observation predominate, is found gradually to collect from the disorderly information, certain impressions, which, in time, become more distinct, until at last an idea is obtained which can be laid hold of. Whether this idea be false or true, is of little consequence; it is in any case enough to give order to the museum of his mind, and for the first time that mind may be said to be informed. Every thing is examined under the influence of this idea, which may change constantly, as it is constantly seen to be unsuited to the facts, but every change may be a progress; and even if from false to false, is nevertheless a constant gain, so much false being left behind. The dreaming age of chemistry had lasted long; the minds occupied had been satisfied either with the more poetical observation of nature, or with the mystic and mysterious; an idea had entered into the mind of the chemists of the age, that some exact explanation could be obtained, and so we find them hunting it down from point to point with acuteness, energy, and hope, ever increasing as the object seemed rapidly to be approached. But how was this idea first attained? How does the first idea rise in the mind of the individual enabling him to see order and beauty in all the shapeless learning that he had been amassing? It is a progress of mind which is not for us to discuss here. The requisite for it is intellectual energy, and how that has been aroused in modern Europe, many great writers have done their best to show. It was by a combination of many great causes, natural stages in the education of the species. We are surprised to find that the finest writers of the world should have existed when thinkers in science were scarcely at the rudiments; but we see constantly that the highest literary powers may be unfit to comprehend any scientific truth firmly, and we see, even among the scientific world, that some branches have still scarcely begun their independent life. The world cultivated poetry and eloquence, and attained the highest stage that we know, when no law in the mixed sciences could be rigidly and certainly stated. The writers, quoted in this chapter, go over again much of the same theoretical ground traversed before, but frequently with much less philosophical grasp than the ancients, and much less largeness of conception, although they frequently gain by what seems the littleness of their ideas, and the smallness of their aims. They are contented to speculate on an acid, instead of the formation of a world, whilst those who take a wider ground, such as Newton and Boscovich, can only be said, as far as our subject is concerned, to reproduce and improve earlier philosophies.

The tendency from this time is to increase the number of bodies, which, if not at first called elements, are at least treated as such. This is a needful step towards the chemical theory of matter in its present stage.

Glauber, although a believer in the general transmutability of substances, did much to help forward the notion of distinct elements in his observations on the affinity of bodies. He explains the evolution of ammoniacal gas from sal ammoniac by a fixed alkali, showing that he understood well combination and decomposition, the stumbling-block of so many, and the introduction to definite compounds. "But that a spirit is distilled off by the addition of fixed salts; the reason is that fixed salts are contrary to acid salts, and if they get the upper hand do kill the same, and rob them of their strength, whereby those things which are mixed with them are freed from their bond, and so it falls out here with salt armoniack, that when by addition of a vegetable fixed salt, the acidity of the salt armoniack is killed; the salt of urine, which formerly was bound therewith, gets its former freedom and strength, and being sublimed turns into a spirit." *

Boyle, who attacked alchemy, and may almost be said to have begun modern chemistry, or rather let us say the transition period, seems to have had one of the clearest, most straightforward, and most common-sense methods of viewing

* Page 49. The Works of the Highly Experienced and Famous Chymist, John Rudolph Glauber. London, 1639.

phenomena, of any of his period. One feels, on reading his works, that on another subject he might have written so as to be even now and at all times read with delight, whereas, in the dangerous and difficult fields of chemistry, he has only left matter serving as landmarks, to shew us the way in which the mind has been obliged to wander in search of truth.

Boyle says, when treating of the "origin of form and qualities," * "There is one universal matter common to all bodies, an extended, divisible, and impenetrable substance."

And in the "Sceptical Chemist." † "But the Aristotelian hypothesis (*i.e.*, of the four elements) is not comparable to the mechanic doctrine of the bulk and figure of the smallest parts of matter, for from these more universal and fruitful principles of the elementary matter, may spring a great variety of textures, upon whose account a multitude of compound bodies might greatly differ from one another." In p. 282:—"Now if it be true, as 'tis probable, that compound bodies differ from one another, in nothing but the various textures, resulting from the magnitude, shape, motion, and arrangement of their small parts, it will not be irrational to conceive that one and the same particle of universal matter, may by various alterations and contextures be brought to deserve the name sometimes of a sulphureous, and sometimes of a terrestrial or aqueous body."

He attacks severely the four elements, the three elements, and the five elements, and justly complains of the "intolerable ambiguity" of the writers on chemistry, and "their playing upon words," as their mode of using salt, sulphur, and mercury decidedly is.

To continue from Boyle. "It seems probable that at the first production of mixt bodies, the universal matter whereof they consist was actually divided into little particles of several

* Vol. I., p. 197. The philosophical works of the Hon. Robert Boyle, Esq. Abridged, methodized, &c., by Peter Shaw, M.D. 2nd edition, 1738.

† Vol. III., p. 206.

sizes and shapes variously moved. * * * * 'Tis also possible, that of these minute particles many of the smallest and contiguous ones were associated into minute masses, and by their coalitions constituted such numerous little primary concretions, as were not easily separable into the particles that compose them." *

† "And indeed if we consider how far the bare change of texture, whether made by art or nature, can go, in producing such new qualities, in the same parcel of matter; and how many inanimate bodies we know to be denominated and distinguished, not so much by any imaginary substantial form, as by the aggregate of these qualities; and that the variation of figure, size, motion, situation, or connection of the corpuscles, whereof any of these bodies is composed may alter the fabric of it; we shall have cause to suspect, that there is no need that nature should always have elements provided, whereof to compose mixed bodies; and that it is not so easy as chymists and others have hitherto imagined, to discern which among the many different substances, without any extraordinary skill, to be obtained from the same portion of matter, ought, exclusive of the rest, to be esteemed its elementary ingredients; much less to determine what primogeneal and simple bodies conspired together to compose it." * * * *

"Our experiment affords us a considerable argument in favour of that part of the mechanical hypothesis which teaches inanimate bodies to differ from one another, but in the magnitude, shape, motion, texture, and, in a word, the mechanical properties of the minute parts they consist of." Page 360. At the end of the chemical doctrine of qualities, p. 441,—"In short, these hypotheses greatly hinder the progress of human knowledge, that introduce morals and politics into philosophy, where all things are transacted according to mechanical laws."

* Vol. III., p. 263. † Page 350.

The doctrine of elementary atoms, either in the sense of indivisible or merely undivided, leads us to suppose all things formed of one substance only; and this seems to have been the notion of those who have held it up as a solution of the question, although generally it has not been carried out far enough to arrive at the difficulties. If all the elementary bodies are the same as this notion supposes, every change is made by change of position of the particles. This is the early doctrine again.

In speaking of the atoms, Boyle says, "little particles of various sizes and shapes variously moved." This "variously moved" introduces a force; it is the introduction of various powers and qualities, and, consequently, of what we term elements. This shews the difficulty of obtaining correct language, when Boyle, whose object was to be accurate, seemed to have overlooked this in those few but important words.

I may quote the graphic words of Boyle, shewing what confused ideas reigned on the nature of combination. "Helmont we know, too, asserts that all mixed bodies spring from one element; and that vegetables, animals, marcasites, stones, metals, &c., are materially but simple water, disguized into these various forms by the plastic virtue of their seeds."* "Aristotle tells us, that if a drop of wine be put into ten thousand measures of water, the wine being overpowered by so vast a quantity of water, will be turned into it. But if this doctrine were true, one might hope, by melting a mass of gold and silver, and by but casting into it lead and antimony, grain after grain, we might, at pleasure, within a reasonable compass of time, turn what quantity we desired of the ignoble into the noble metals."†

Lemery, born about twenty years after Boyle, felt, but not very clearly I imagine, some of these difficulties, as he

* Sceptical Chymist. Vol. III., p. 284. † Page 289.

says, "the first principle that can be admitted for the composition of mixts is an universal spirit, which being diffused through all the world produces different things according to the different matrixes or pores of the earth in which it settles. But, because this principle is a little metaphysical, and falls not under our senses, it will be fit to establish some sensible ones." *

Here, again, is a notion that matter is one, but with an addition that the different properties of its parts are caused by abstract qualities; here there is nothing said either of particles or transformations. It would seem as if matter existed in some condition as one, and received qualities from a spirit which had the power of giving various qualities. We see no variety of agents, but one agent; so that a will and design are constantly required to be present.

This is manifestly a backward step from Boyle, who was not well followed up in his own direction.

Mayow, of Oxford, wrote in 1673 definite opinions as to the combination of acid and alkali, shewing that sulphuric acid separated nitric from saltpetre, and formed vitriolated tartar, or sulphate of potash, that therefore the nitric acid was not destroyed when it united with potash, seeing that it could be separated again by the sulphuric acid. Although we cannot separate the sulphuric acid from the potash, "it is not because these salts have mutually destroyed each other, but because there is nothing in nature which unites with each more firmly than they do among themselves."† He also clearly shews that alkalies throw down metals from their solutions.

"These important observations of Mayow were continued still further by Geoffroy Senior. He considered the order in

* Page 2. Lemery's Course of Chymistry. Translated by Walter Harris, M.D. London, 1686.

† Tractatus quinque Medico Physici, 1674. Chapter, "De salium contrariorum congressu et precipitatione."

which bodies separated each other from a given body as constant. Thus metals are separated from acids by the absorbent earths; the absorbent earths are separated from acids by volatile alkalies, while volatile alkalies are separated by the fixed alkalies. He drew up, in consequence, the following tables, exhibiting the order in which bodies separate each other from a given substance. At the head of each column is written the name of the substance with which the bodies, enumerated in the column, combine. Below it are arranged all the bodies capable of uniting with it. That which separates all others is placed uppermost, and that which is separated by all the others is placed undermost, and the others in the order of their separations. Thus, in the first column, the fixed alkalies separate all the bodies below them from the acids. The volatile acids separate all except the fixed alkalies. The absorbent earths separate the metals, and the metals are separated by all the other bodies in the column."

ACIDS.	MURIATIC ACID.	NITRIC ACID.	SULPHURIC ACID.	ABSORBENT EARTHS.
fixed alkalies	tin	iron	phlogiston *	sulphuric acid
volatile alkalies	antimony	copper	fixed alkalies	nitric acid
absorbent earth	copper	lead	volatile alkalies	muriatic acid
metals	silver	mercury	absorbent earth	
	mercury	silver	iron	
	gold		copper	
			silver	

FIXED ALKALIES.	VOLATILE ALKALIES.	METALS.	SULPHUR.	MERCURY.
sulphuric acid	sulphuric acid	muriatic acid	fixed alkalies	gold
nitric acid	nitric acid	sulphuric acid	iron	silver
muriatic acid	muriatic acid	nitric acid	copper	lead
acetic acid		acetic acid	lead	copper
sulphur			silver	zinc
			antimony	antimony
			mercury	
			gold	

* Geoffroy's expression is *principe huileux*. Kopp's Geschichte der Chemie.

LEAD.	COPPER.	SILVER.	IRON.	ANTIMONY.	WATER.
silver	mercury	lead	antimony	iron	alcohol
copper	calamine	copper	silver, copper, lead	silver, copper, lead	salt

This, on Geoffroy, is taken from Thomson's Chemistry, edition of 1817. The tables are in all the older chemical books. I have not seen the original.

Although these tables were extended and improved by Gellert and Sage, they were not altered in principle until the middle of last century. Although every thing is indefinite here, that is, unfixed by numbers representing either weight or measure, a certain order is given to the substances known to chemists, a classification is produced, certain laws are laid down, and some relation of one body to another is given, at the same time bodies begin to get a definite character and position, and are not mixed up in a confused mass together, capable at any moment of playing the most fantastic tricks with the operators, and converting the chemist from a naturalist to a magician.

In speaking of atoms we must not confine ourselves to bodies that cannot be divided; Newton and others being satisfied with such as are not usually divided. Newton's well known words are these:—" It seems probable to me that God in the beginning formed matter in solid, massy, hard, impenetrable, moveable particles, of such sizes and figures, and with such other properties, and in such proportion to space, as most conduced to the end for which He formed them; and that these primitive particles being solid, are incomparably harder than any porous bodies compounded of them; even so very hard as never to wear, or break in pieces; no ordinary power being able to divide what God himself made one in the first creation. While these particles continue entire, they may compose bodies of one and the same nature and texture in all ages; but should they wear away, or break in pieces, the nature of things depending on them would be changed. Water and earth, composed of old worn particles and fragments of particles, would not be of the same nature and texture now with water and earth composed of entire particles in the beginning. And, therefore, that nature may be lasting, the changes of corporeal things are to be placed only in the various separations and new associations and motions of these permanent particles; compound bodies being apt to break, not in the midst of solid particles, but where those particles are laid together, and only touch in a few points."*

This is the perfection of the mechanical theory, unaided by chemistry. It provides for a diversity of particles in size, shape, and other properties, as well as diversity of arrangement, and as this explains what may be called the external phenomena of combination, it is, to a certain stage, probably the correct one. Our difficulties arise when we think on the constitution and mode of combination of these particles. On the first Newton scarcely allows any argument, on the second he has not said sufficient to satisfy the demands of chemistry.

Dr. Shaw, although not a great discoverer, was a clear-headed man, and his chemical lectures, in 1731, 1732, and 1733, are amongst the first in which we see the matter plainly and practically handled, following up the instructions of Boyle, which he carefully edited. †

At p. 146, he says:—" We must, therefore, observe that the more intelligent of the modern chemists do not understand by principles, those original particles of matter of which all bodies are, by the mathematical and mechanical philosophers, supposed to exist. Those particles remain undiscernible to

* Newton's Optics, near the end.

† Chemical Lectures, publickly read at London in the years 1731 and 1732, and at Scarborough, in 1733, for the Improvement of Arts, Trades, and Natural Philosophy, by Peter Shaw, M.D., F.R.S. Second edition, 1755.

the sense, though assisted with the most finished instruments, nor have their figures and original differences been determined by a just induction. Leaving, therefore, to other philosophers the sublimer disquisition of primary corpuscles or atoms, of which many bodies or worlds have been formed by the fancy, genuine chemistry contents itself with grosser principles, which are evident to the sense, and known to produce effects in the way of corporeal instruments."

As it is clear that matter has various appearances, so if there be only one matter, it must be capable of putting on various forms. This variation of forms has been given as an innate power of matter, and if we extend this idea, we come to the conclusion, more or less clearly expressed by many writers, from the earliest times, that matter itself has no qualities; but that we perceive only the qualities it has put on. Quality means, then, the true matter, or sensible thing which we see. A great many metaphysical, as well as physical difficulties have been removed by allowing a greater number of elements, leaving the difficulties to be solved of a much more profound character.

The remarkable position given to character, making it play the part of matter, is seen well in Hooke's works.*

"I conceive the whole of realities that in any way affect our senses, to be body and motion. By body I conceive nothing else but a reality that has extension every way, positive and immutable; not as to figure, but as to quality; and that the body, as body, is the same, whatever figure it be of: as a quart of water is a quart of water, or a certain quantity of body, though contained in a globe, cylinder, cone, cube, quart-pot, or any other figured containing vessel; and as body, it is indifferent to receive any figure whatever; nor has it more extension in the one than in the other vessel, nor can it have

* The Posthumous Works of Robert Hooke, M.D., F.R.S. Folio. 1705. Pages 171-2. He died in 1703.

less; nor is it more essentially a body, when solid, as ice, than when fluid; that is, the minims of it are equally disposed to motion or rest in position to each other; and therefore body, as body, may as well be, or be supposed to be, indefinitely fluid as definitely solid; and, consequently, there is no necessity to suppose atoms, or any determinate part of body perfectly solid, or such whose parts are incapable of changing position one to another; since, as I conceive, the essence of body is only determinate extension, or a power of being unalterable of such a quantity, and not a power of being and continuing of a determinate quantity and a determinate figure, which the anatomists (or atomists) suppose. These, I conceive, the two powers or principles of the world, to wit, *body* and *motion*, uniformity of motion making a solid, and difformity of the motion of the parts making a fluid, as I shall prove more at large by and by."

* * * * "As for *matter*, that I conceive in its essence to be immutable, and its essence being expatiation determinate, it cannot be altered in its quantity, either by condensation or rarefaction; that is, there cannot be more or less of that power or reality, whatever it be, within the same expatiation or content; but every equal expatiation contains, is filled, or is an equal quantity of *materia;* and the densest or heaviest, or most powerful body in the world contains no more materia than that which we conceive to be the rarest, thinnest, lightest, or least powerful body of all; as gold for instance, and *æther*, or the substance that fills the cavity of an exhausted vessel, or cavity of the glass of a barometer above the quicksilver. Nay, as I shall afterwards prove, this cavity is more full, or a more dense body of æther, in the common sense or acceptation of the word, than the gold is of gold, bulk for bulk; and that because the one, viz., the mass of æther, is all æther; but the mass of gold, which we conceive, is not all gold, but there is an intermixture, and that vastly more than is commonly supposed, of æther with it; so that

vacuity, as it is commonly thought, or erroneously supposed, is a more dense body than the gold as gold. But if we consider the whole content of the one with that of the other, within the same or equal quantity of expatiation, then are they both equally containing the *materia* or body."

This argument will not appear so conclusive to his readers, but it serves well to shew how unsettled were the opinions on matter, when a man of Hooke's high standing spoke in this manner. He could not believe in epicurean atoms as he called them, and he had as little faith in "the four elements, the three chemical principles, magnetism, sympathy, fermentation, alkaly, and acid, and divers other chimeras." The quotation is by no means intended as a specimen of Hooke as a philosopher. Aristotle's aether and the abstract ideas of Plato are still perceptible here in confusing the reason and obliterating the observation.

When a definite form is given to the minutest particles of bodies, it leads to their indivisibility by a very easy reasoning, but the moment these ideas become vague, or if a prime matter is allowed, taking various forms, or a few elements changing into each other, essentially the same things, then infinite divisibility is more likely to lay hold of the imagination.

Des Cartes, speaking of the theory of Democritus, says, "it is rejected because he supposed the existence of indivisible corpuscles, because there was a vacuum around each, which could be demonstrated impossible, and because there was weight given to each, whereas no body has it when taken alone, its existence depends on the situation and motion of other bodies."* But he thinks also that natural things might have been made in various ways. We must explain the phenomena as we best can. "Ita non dubium est, quin summus rerum opifex, omnia illa quae videmus, pluribus diversis modis potuerit efficere."† He also retained the four elements.

* Renati Des Cartes, Principia Philosophiae. Page 219. Amsterdam, 1677.

† Prop. ccii.

We find Sennertus, about the same time, expounding Aristotle's opinions on the elements, and looking to the authority of character more than experiment, but he does not give a fifth element such as Aristotle's ether. The argument he gives that there are only four is curious, and another of the interminable varieties of method in which vagueness thinks. "There are two right motions, one from the mean, the other to the mean, there will, therefore, be an equal quantity of simple bodies, subject to these two simple motions, one which absolutely is heavy, called earth; another absolutely is light, called fire. But because nature wishes the world to be one, but contrary extremes cannot constitute one, she always couples the extremes by means (per media), and connects the last of the superior kind with the first of an inferior. This mean is therefore required. But this cannot be *one*. Because, if so, it would occupy the mean place between the extremes, or between the centre and circumference, and so no right motion (rectus motus) could be given to it. For it could neither be moved to the middle, nor from the middle, nor could it be called more or less heavy. It is necessary, therefore, to have two means, one light, and which may be moved from the middle upward, in respect of which it is heavy, and is called air; the other heavy, and tending to the middle, in respect of which it is light, which is called water. There are, therefore, four elements, fire, air, water, earth. There cannot be a fifth for the same cause, that there cannot be only one mean. But if any one will desire to establish five, his senses and experience disprove it." *

He endeavours to establish a difference between the pure fire and the common drudge of life, and quotes Scaliger to support him, but in neither of them do we find any distinct ideas of what these differences consist. We are confounded by the great part which hot, cold, moist, and dry, have to play, and it

* Danielis Sennerti *Vratislaviensis* Epitome Naturalis Scientiae. Oxoniae, 1632. Page 185.

never occurs to them that hot air is air and heat, and that moist air is air and moisture, but they are qualities put on by air at will, or at command, by accident, by growth, by motion, or by life. The qualities of the elements are not substantial forms, but accidents; the original first matter forms the elements which put on various forms, and so perform everything that elements have the power of accomplishing. The changes are performed by generation, fermentation, and corruption.

Baumé, a believer in the four elements and phlogiston, in 1773, says, that it is of no advantage to consult the ancient chemists, they are so ambiguous; and says: *—

"Is it not probable that nature combines the elements in a direct manner by twos and twos, and by threes and threes, by means absolutely unknown to us? If these simple combinations exist, they will be secondary principles, or principles made from principles (principes principiés), of which nature makes use to form compound bodies. Knowledge, on this subject, entirely fails. We have no information on the immediate combinations of the four elements; we only know that they have such a disposition to mix, that it is impossible to have them perfectly pure and isolated from each other."

We have here an opinion freed from the shackles of salt and sulphur, but not beyond the early times, and I think not so penetrating, although as practical as Roger Bacon's.

It will be interesting here to quote Bishop Watson, a man who brought into chemistry more common sense and less pretension than many less penetrating observers. Speaking of elements, he says, "by chemical elements, which are the last products of chemical analysis, we are to understand, not very minute indivisible particles of matter, but the simple homegeneal parts of bodies which are not capable, as far as our experience teaches us, of any farther resolution or division, except in a mechanical sense, into

* Page 119, Vol. I., of "Chymie Expérimentale et Raisonnée."

similar parts, less and less, without end, as water into vapour, more or less, subtile and attenuated. Aristotle and his followers esteemed earth, air, fire, and water, to be elements, simple and uniform in their several kinds, essentially distinct and utterly incapable of being converted into one another, yet easily uniting together, and by their different arrangements, and properties and mixtures, composing every body in the universe. Many modern chemists have adopted this idea; others have increased the number of elements, by adding a saline principle; others have contended that some of these elements, air and fire for instance, are themselves compound bodies; and others, lastly, are persuaded, that there is only one elementary homogeneal principle, and that all the varieties of bodies, as well as of what are most commonly esteemed elements, ought to be attributed to the different magnitudes and figures of the particles composing them; and as the component parts of water or air, or any other body, are by no means supposed to be elementary particles of matter, but to be made up of different numbers of elementary particles, arranged in different forms, it may be thought probable, that mechanical causes may either diminish or augment the number, or change the disposition of the particles, and thus effect the several varieties observable in nature.

"It would be improper in this place to enlarge on a subject, concerning which both ancient and modern philosophers have been so much divided in opinion. Their great diversity of sentiment may suggest a suspicion that the full comprehension of it does not fall within the reach of the human understanding. The following observation may, perhaps, tend a little to illustrate this matter. Let us suppose that this terraqueous globe was not surrounded with any air or atmosphere, and that by an approach to the sun, or an increase of subterranean fires, by some means or other it should become exposed to a heat four times greater than the medium heat of summer, which we may reckon to be about 60 degrees of Fahrenheit's thermometer; then would an atmosphere be quickly formed around it, all the water on its surface, most of the juices of plants and animals, and a great variety of mineral particles, would be raised up in vapours and exhalations, and whilst the heat continued would be kept suspended in an elastic state, and constitute an atmosphere analogous, as it may reasonably be imagined, to the chaotic state of our present atmosphere, only differing from it in this, that it would require a greater degree of heat in order to keep the particles of matter from coalescing into one heterogeneous mass. Again, in the present state of the atmosphere, suppose that a great degree of cold should continue unabated for any length of time, all the water on the surface of the earth would be changed into a solid transparent stone, which might be dug out of its quarry, and then employed in building as well as marble, or any other species of stone; all the particles of air would be brought closer together; some of them which were the least elastic would be reunited: and imagining the cold to be indefinitely increased, what reason can there be against supposing that the whole atmosphere would be reduced into a solid state, forming a heterogeneous crust on the surface of the earth; the thickness of this crust, supposing it to be as dense as marble, would be about four yards? It will easily be understood, that water, and air, and earth are, upon this hypothesis, but variations of the same element introduced by heat.

"That the atmosphere which surrounds the earth was originally formed from the chaotic mass, by having the more subtile parts of which that mass consisted, elevated and put into an elastic state by means of heat, seems not altogether improbable. We find the atmosphere or firmament immediately succeeding the formation of light; now, if the effect of that light was heat, be the form or matter of it what you please, then would such particles of the shapeless jumble as were capable of being evaporated with that degree of heat, be elevated in an elastic state, and a division or separation would be made in the midst of the great abyss, between the waters which were of a nature subtle enough to be converted by that degree of heat into an elastic fluid, constituting the firmament or atmosphere, and the waters which could not be evaporated in that degree of heat, but still remained covering the surface of the globe, being not collected into one place, that the dry land might appear, till the third day. This notion of the atmosphere and its formation, seems to be conformable enough to Newton's opinion expressed in his letter to Mr. Boyle. "I conceive the confused mass of vapours, air, and exhalations, which we call the atmosphere, to be nothing else but the particles of all sorts of bodies of which the earth consists, separated from one another and kept at a distance by the said principle," a principle of repulsion." *

Here we have the one element again or matter treated simply as matter generally, in the same way in which metaphysicians have dealt with it, leaving heat and, probably as by others expressed, some general plastic power to form all the modifications. This is much too indefinite for science, and it is surprising that it should have been deemed sufficient even in Bishop Watson's time. There is no attempt to examine the degree of heat which may convert a solid body into a permanent gaseous one, and no difficulty perceived in the persistence of the atmosphere in a gaseous state whilst exposed to no more cold than the solid or liquid substances around us, or the persistence of the water in a liquid state, or the permanency of the earth, why it did not occasionally send off rocks into vapour or rarefy them into air; nor is there any attempt to find what strange powers may cause the diversity of structure. Although not expressed, these words were evidently

* Chemical Essays. By R. Watson, D.D., F.R.S., &c. Vol. I., p. 100. 4th Edition. London, 1787.

written under the idea that the powers of nature worked rather arbitrarily and capriciously; how, is not distinctly seen. When the mixture of pure elements is spoken of as forming all bodies, there is a reason given clear enough for the changes, and the chief blame attached to those who believed in this is, that they did not obtain the amount of each needed to form any one distinct substance. This proves, as it appears to me, that they held their opinions by a very slender tie.

Boscovich, in his theory of natural philosophy, in 1759, gave the fullest scientific expression of the unsubstantial theory of matter, which may be called the *dynamical*. The book is not common, and not to be found in this neighbourhood. I shall take Dr. Daubeny's description: "He supposes that matter is made up of a number of unextended indivisible points, which, however, never touch each other, owing to the mutual repulsion subsisting between them, so soon as they come within a certain distance of each other; which repulsion increasing gradually in proportion as they are made to approach nearer and nearer, becomes at length too powerful for any force to overcome." *

In this theory we have again revived in another form the idea that all matter comes from a non-material, or what we may call a spiritual force, nor is it easy for us to conceive how it could have any other origin. It is a revival of the doctrine of the mind and of the soul being the origin of matter, or of the early opinion that numbers were the true beginning, or in other words, abstract forces. But we have to do with matter when it is formed, not with its production; and when these unextended points have obtained existence, we are obliged to reason on them as if real. One result, however, affects our subject, that by this theory, matter may cease to be infinitely

* An Introduction to the Atomic Theory. By Charles Daubeny, M.D., F.R.S., &c. Second edition, Oxford, 1850. p. 34.

divisible. It does not of necessity cease to be so, as we may readily believe, in the constancy of these points; but we may also imagine the power dispersing in various directions, and the absolute amount of force existing in one particle to lie diffused throughout the whole world, or any given amount of space.

When the particles are formed, Boscovich deals with them exactly as with matter, formed of indestructible atoms. "Thus he supposes that the points of matter alternately attract and repel each other, according to the distance that separates them, until they either come very close to, or are removed to a comparatively great distance from each other; in the former case they are repelled, in the latter attracted; the former force preventing mutual contact, the latter, which, when considered as acting between the earth and bodies upon it, is no other than gravitation, drawing them all together."

The *Encyclopedie Methodique*,* in an article written by G. Morveau, 1786, may be presumed to give the advanced opinion of chemists. He says,

"Are there really different degrees of saturation of the same salt? or is not the union which it contracts with that portion which exceeds the point of saturation, the effect of super-composition, as in combination with a third foreign body? This is a question which merits all our attention, not only because it is of interest in the general theory of chemical attraction, but rather because it is of importance that we should understand what is the character of the affinity before we attempt to submit it to calculation, or even deduce a satisfactory explanation of the phenomena which depend on it. I confess that the idea of divers degrees of saturation of one body with another appears to me repugnant to all the notions

* Vol. I., p. 560-1. Encyclopedie Methodique. Chymie Pharmacie et Metallurgie. Paris, 1786—1815.

we have hitherto acquired of the method by which combinations are formed. I can readily conceive that the point at which water is saturated with a salt may change according to the temperature, and that cold water may absorb more of the gaseous acid; the increase or diminution of the matter of heat in the solvent changing the respective disposition, the density, and perhaps the figure of the molecules, it is not astonishing that the attractive force should be modified by these changes; that they should result in a contact more or less perfect, and that the power of affinity should be able by this means to protect from *(the influence of)* the law of gravitation a greater quantity of matter, in the one case, than in the other; but we have nothing of this kind in the hypothesis under consideration; the circumstances are the same; the point of saturation cannot change, because it is the effect of a cause which does not change.

"To make this clear, let us ask what is saturation? Every chemist will say that he understands by it the condition in which a compound is, when neither of its constituents can receive or retain in combination a greater quantity of the other. Such is the rational and necessary acceptation of the word saturation, otherwise it becomes void of sense; then to suppose, preserving this acceptation, that a substance may be saturated with different quantities of the same substance, is really to affirm two contraries."

"* * * * Now, when a new quantity of any of their principles is added to one of these perfectly neutral salts, there is nothing to prevent an attraction between them, and even in a degree capable of producing solution, combination, or affinity; but it must be remarked that this affinity is not that of an acid to a base, or a base to an acid, but of the neutral compound, with the portion that was added; whence it follows; 1st, that it has no effect on the previous composition which remains in its integrity, as if the neutral salt were supercompounded with a foreign body; 2nd, that the point of saturation is not changed; and 3rd, that the power which unites this portion which is added to the neutral salt, may be much weaker than that which unites the same substance to the same, at the point of saturation, without causing a contradiction."

Page 566. "I shall terminate this section by a short résumé of the principal characters, which are capable of forming a methodic division of affinities, and which may have been lost sight of in the course of the preceding discussions.

"1st. *Two bodies of the same nature*, whether simple or compound, may unite and form a third, as homogeneous as either of the two before the union. This is called *affinity of aggregation*.

"2nd. *Two bodies of a different nature*, simple or compound, may unite without undergoing any change in their first composition, if they are compound. This is called *affinity of composition*.

"Two out of three bodies may shew a preference and combine, leaving the third at liberty; these bodies may either be simple or compound, provided their composition does not change, and they are found in a condition favourable to contact. This is still affinity of composition; whether the three bodies have been put separately into the mixture, or two been previously united, whilst the superior affinity of the third has destroyed their union, by what is called precipitation.

"*Three or more bodies*, exposed to contact, may unite in such a manner as to form only one homogeneous mass. This is nevertheless affinity of composition, although two only unite at first, and a third is added to the first two, and so on successively.

"3rd. *Two bodies* incapable of combining, become so when one or the other has either been decomposed or supercomposed.

In this case, the affinity of composition, which produces the union, takes the name of *disposing affinity*.

"4th. *Two or more compounds* being placed in circumstances, suitable for bringing into play the respective affinities of their component parts; either there is a change and new products formed without our being able to determine which is the most powerful affinity under which they act, or the first composition remains, contrary to the order indicated by the superior affinity of the principle of one of the component parts to the affinity of the other. In these two cases we say that they are not the relations of affinity of one body to another, *but affinities of concourse*, otherwise called *double affinities;* in a word, the sum of all the united affinities which are needed to explain these phenomena.

"Two bodies being put in contact, the compound which results is supercompounded, or united with an excess of one of the principles. This tendency to supercomposition, is sometimes so strong, that when the least affinity of a third body interferes with it, the proportions of the first compound are changed, and the neutral state destroyed. This tendency may cause the production of crystals, with excess of base, in an acid liquor. To distinguish this force, we shall call it the affinity of a compound for an excess of one of its constituents, or for shortness, *affinity of excess*, which will be enough to recall the idea when it has been well grasped. However paradoxical some of these propositions may appear, I have no fear of their being called in question after the proofs I have given; and if they are well founded, it will readily be granted that they ought to form one of the most important elements in the calculation of affinities."

Here now we have what Morveau has given, as the most certain of the laws of affinities known among chemists at his time. He adds, however, that they certainly scarcely deserve the title. The following are additions :—

* "1st. There is no chemical union, if one of the bodies be not sufficiently fluid to allow the molecules to obey the law of affinity, which carries them from (mere) proximity to (actual) contact.

"2nd. Affinity takes place only among the smallest integral molecules of a body.

"3rd. From the affinity of one substance with another, we cannot know the affinity of a compound; we cannot know the affinity of a compound of one of these substances, with one or the other in excess.

"4th. The affinity of composition has no efficacy, unless it can overcome the affinity of cohesion.

"5th. Two or more bodies, which unite by the affinity of composition, form a substance which has new properties distinct from those which belong to each of the bodies before combination.

"6th. To give effect to affinities, a particular temperature is necessary, which renders the action either slow or rapid, invalid or efficacious."

These feeble and incipient attempts at laws may surely surprise us, written as they are almost within the memory of some of the living. There was at this time an attempt to measure the force of affinities as the only method of obtaining results, and there was not yet seen the absolute necessity of having bodies kept uniform by a constant and absolutely similar composition, much less to bring this absoluteness of composition under the forms of natural law, and we might add also, logical necessity.

It should be noticed that the remarks about saturation shew that compound proportion was not in the least degree understood in the present sense. Its possibility is even denied.

The aim was to find the strength of affinity. When it is

* Remarks in italics on the various laws, from p. 567 onwards.

said that we cannot find the affinity of a compound from knowing that of its parts, it refers to affinity dynamically; no hint is given as to the knowledge of this amount quantitatively. Proportion generally may be said to be excluded.

Although these opinions give a pretty fair idea of the state of the chemical mind at the several dates, we must return to an earlier date again to follow up another direction.

CHAPTER VII.

PHLOGISTON PERIOD AND PROGRESS OF THE BALANCE.

In the 17th century innovations were beginning in chemistry, as we have already seen, but as usual, these did not all take one direction. Van Helmont put his little *archaeus*, a kind of intelligent agent, but with less independence than that of Paracelsus, into the stomach, to do the work which he could find no way of accomplishing by merely physical means. Thus things began a new mystical direction. Becher, in the *Physicae Subterraneae*, ridicules his archaeus and chimeras, and the whole host of "impudent chemists" also, who assert that they obtain salt, sulphur, and mercury, from all bodies, even animals and vegetables. He does not hesitate to call these the greatest falsehoods. He calls "elements the genuine and true things of which bodies consist, and from which others are made and prepared." * But as he held on by the four elements, we are not able to find in him much material.

He had the merit of raising inquiry in a high degree, and of bringing forward his great admirer, Stahl, who introduced phlogiston. In this chapter we have a class of men who have made another advance in experimenting, and whose works are the first which living chemists can, without difficulty, peruse. The advent of oxygen into science was preceded by a century of vague prophesyings. The use of the balance was becoming general, but men had no idea of the accuracy with which nature weighed, although they had long used the proper principle of making the earth the arbiter, by trying which side of the scale she drew most willingly towards her. They

* Phys. Sub. Lib. i., sect. iii., cap. i., No. 12.

had no idea of the fineness of her touch, and her absolute refusal to make any allowance for inaccuracy in the construction of instruments. It was not even known that all bodies could be compared by their weights; why should they not as well be known by their lightness? This plan had its fair trial. By a curious circle of reasoning, it was decided, that what we call oxygen, which makes an oxide, or calx of a metal, was sulphur; afterwards it was the principle of combustion; not such an erroneous idea. Now oxides or earths were, *of course*, simple bodies; when they were reduced to metals in the fire, they combined with phlogiston; they became lighter. Therefore phlogiston had the principle of lightness in it. The rule generally is, that we should begin wrong. We now say the metal is simple, and by uniting with oxygen, it becomes a compound, and is heavier. As the metal burns and gives out heat, they said it gives out its phlogiston, and loses its principle of lightness.* Stahl calls it sulphur. This would scarcely come under our view had it not been the cause of so many inquiries in the same direction, as to bring about a result, derived from an analysis of all the oxides, and a careful comparison of the weight of the metals, with the weight of the oxides, whether produced by combustion or oxidation in the fire, or precipitated from their acid solutions. Even this strange theory tended in the right direction, although at first threatening to take a mystical course. We could scarcely have anticipated this difficulty of proving that all bodies have weight and not lightness, but our forefathers encountered it, and it may yet come to the struggle again, renewed in a higher form, when we have to deal with those physical existences, now called *imponderables*.

I am not aware that any one went into the subject with care before Bergman. He may be said to have introduced modern analysis. Before him analyses were not superior to

* p. 277. "Traité de Soufre" Traduit de l'Allemand de Stahl. Paris, 1767.

those speculations about the constitution of bodies which in former chapters have been passed over. I may indeed cite here Roger Bacon's syntheses of bodies from the four elements, as the earliest examples of an endeavour to shew how so many bodies can be formed from few elements, and on the other side, as the fullest example I know of early analysis, and perhaps the very first in which numbers are used in connection with elements. They are intellectual strivings after quantitative analysis.

"There is, therefore, one different kind where fire and air are greater; 2ndly, where fire and air are less; 3rd, where fire and water are greater; 4th, where fire and water are less; 5th, where fire and earth are greater; 6th, where fire and earth are less; 7th, where air and water are greater; 8th, where air and water are less; 9th, where air and earth are greater; 10th, where air and earth are less; 11th, where water and earth are greater; 12th, where water and earth are less; and so you have two diversities. Next you have three diversities; 1st, where fire, air, and water are greater; 2nd, where fire, air, and water are less; 3rd, where air, water, and earth are greater; 4th, where fire, water, and earth are greater; 5th, where air, water, and earth are less; 6th, where fire, water, and earth are less; and in this manner, if you divide those methods, you obtain from the first 16, from the second 64, from the third 47, from the fourth 18, in all 145. I will now speak of the fourth diversity, fire much, air less, water much, earth less; second, air much, water less, earth less, fire less, and so being ingenious, you may draw out all these diversities to light."

A manuscript copy of Dr. Cullen's lectures in 1762-3 in the laboratory of Owens College, Manchester, from the late Dr. Henry's library, mentions four elements, which, by simple combination, could be formed into seven, but any proportionate combination to account for the number in nature, is not given.

These lectures shew him to have been an exceedingly clear and rational expounder of science. With good common sense he waits for more knowledge when science fails, fully shewing why he became famous, although he published very little. Reasoning on the state of things at the time, he says, "It appears, then, that we know of no physical element, nor any chemical principle, nor are we acquainted with any body which has fixed and permanent qualities."

He afterwards adds, "Having laid down and demonstrated this fundamental proposition, viz., that the changes of the qualities of bodies are all of them produced by combination or separation, I now proceed to inform you that combination depends upon *attraction*, that is, the attraction of cohesion, whereby the small particles of bodies very near each other are disposed to approach, and in a certain contiguity to remain coherent together."

He then goes on to explain simple elective attraction and double elective attraction by diagrams, like those below, where the lines ought to be drawn straight from C to B, and from A to D. This appears to be earlier than Bergman, who at that time had published nothing on chemistry. I can find no internal evidence of their being written later than they profess to be, the binding itself being old.

Dr. Cullen was professor of chemistry at Glasgow, and Dr. Black attended his lectures, before being appointed his successor, on the removal of Dr. Cullen to Edinburgh, in 1756. In the Annals of Philosophy, Vol. III., p. 554, Dr. Thomson says:—"My knowledge of Dr. Cullen's opinions was derived from the late Professor Robison, of Edinburgh, who had the means of information, and, as he was a particular friend and great admirer of Black, is entitled to credit. Now, he informed me that Dr. Black's explanation of double decompositions, which he annually gave in his class, had been originally broached by Dr. Cullen. This was the circumstance that induced me to broach Dr. Cullen's name along with Black and Bergman.

"* * * As to Dr. Black, I consider myself as acquainted with his opinions, because I attended his lectures; and there are thousands in Great Britain who did the same, and who cannot but recollect the facts that I shall state. Dr. Black taught that bodies combine in definite proportions, and he explained double decomposition by means of diagrams, not, indeed, the same as those of Mr. Higgins, but much simpler and more elegant. I have been informed by Prof. Robison, that he employed these diagrams from the very beginning of his career, as a professor. One of them is given in page 554, Vol. I., of his printed lectures. I have no doubt that all similar diagrams, published in London, by Fordyce, &c., were derived from the same source. Now, could the doctrine of definite proportions be taught, and could double decomposition be explained in this way (I quote Dr. Black's explanation), let the bodies A and B be united with a force, 10; and the bodies C and D with a force, 6. Suppose the attraction of A for C to be 8, and that of B for D to be 9, if we mix these bodies, A will unite with C, and B with D. To me they conveyed just as much of the atomic theory as the perusal of Mr. Higgin's book did."

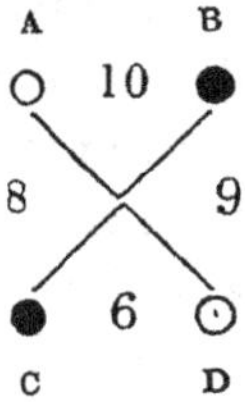

Dr. Robison edited the lectures of Black in 1803, and in a note gives the above diagram and some judicious remarks, shewing, at the same time, that although definite proportion was taken for granted, no general law to account for it had been given.

But the question cannot be as to whether Dr. Cullen discovered the atomic theory, (indeed, this extract might have been brought on somewhat later), but whether Dr. Cullen had so far advanced our knowledge of matter as to be the first who gave out the ideas of single and double elective attractions, such as have been attributed to Bergman.

A note gives a fuller account of Dr. Cullen's views; it was written in the year 1759. Elective attractions were in reality definitely laid down and presupposed in Geoffroy's tables; but the investigation and elaboration was needed.* At present we must consider Dr. Cullen as the first who used the words and explanations in the manner afterwards made so famous by Bergman.

* Note E., p. 45., Cullen's Life, by Thomson, p. 570. Appendix.—The following passages from a letter, written by Dr. Cullen to his friend and former pupil, Dr. George Fordyce, of London, in October, 1759, contains his own statement of his views with regard to double elective attractions. "I must give you the manner of considering the subject, which I fell upon last session, and shall continue to employ as the most easy and simple. I begin with your third and fourth cases, and to these one general rule applies, viz., that when two mixts (compounds) are applied to each other, if in each mixt there is a substance, that from the table of elective attractions, is by itself capable of decomposing the other mixt, the attractions between these substances and the substances they attract in the opposite mixt, must always be greater than the attractions subsisting in the mixts applied to each other; and therefore, &c. Thus, if *nitrum argenti* and common salt are applied to each other, as by the table of elective attractions, the nitric acid in nitrum argenti, is by itself capable of decomposing the other mixt, common salt; and the muriatic acid in common salt is capable of decomposing *nitrum argenti:* the attractions between the nitric acid and the soda, with the attraction of the muriatic acid and the silver must be always greater than the attractions subsisting in the mixts, nitrum argenti and common salt, that were applied to each other. This I illustrate by the diagram adjoined. Let there be two rods intersecting one another, and moveable on a common axis at the point of intersection. At the extremities of each let there be placed substances that have an attraction for each of the substances on the extremities contiguous to them, and let the attractions be expressed by the letters W, X, Y, Z. The rest of the illustration will readily appear from the diagrams.

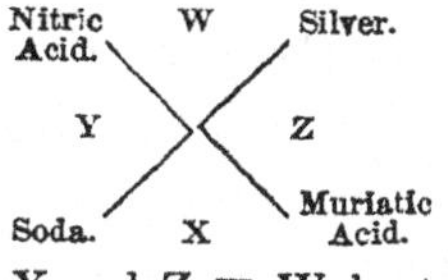

$Y > X$ and $Z > W$ by table.
Ergo $Y + Z > X + W$

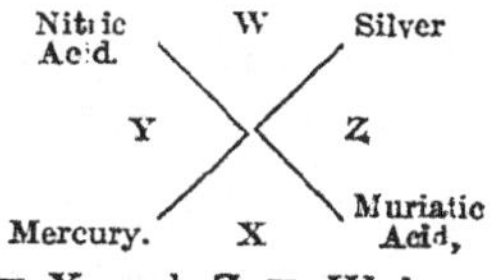

$Y > X$ and $Z > W$ by table.
Ergo $Y + Z > X + W$

You see that the prevailing attractions are here determined from the table of single elective attractions.

We are now come to the only difficulty in the affair of double elective attractions in instance past. To this our general rule does not apply.

There is no doubt that, however these opinions might be at the time floating amongst chemists, the works of Bergman were both the fullest and the most important on this subject. From them I shall give rather long quotations. *

"On the different amount of phlogiston in metals,† he says; calces (oxides) do not displace each other, as experience shows, at least, not in the same order as the metals do. May not therefore the quantity of reducing phlogiston in any metal be determined by a comparison of the weights of the precipitated and the precipitating metal? The following experiments will show the answer, but let us first examine, in a general way, those cases which may possibly occur:—

"Let A be the precipitating metal, m the weight of acid necessary for dissolving 100 of A, x the quantity of reducing phlogiston in 100 of A; B the metal to be precipitated, nm the weight of the solvent mentioned necessary for dissolving 100 B, and y the amount of reducing phlogiston in 100 B. n may be equal to unity, or it may be more or less than unity."

"Let, I., $n = 1$ then $m = nm$."

(In other words, if $n = 1$ the quantity of acid necessary for dissolving the precipitating metal, it will be equal to the quantity necessary for dissolving the precipitated metal.)

"In this case, if $x = y$ there is no difficulty, because the solvent of each dissolves an equal weight, and B is able to take from A as much of the inflammable material as is necessary for its reduction.

See how it comes out when my new scheme is applied to it. Y and Z are by the table of elective attractions each of them less than W, but greater than X. If, therefore, Y and Z are exactly as much greater than X, as they are less than W, the four attractions would be exactly balanced; but if Y and Z exceed in any degree more than they fall short of W, than Y + Z must be greater than W + X.

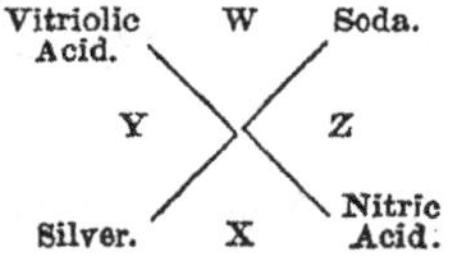

* Torberni Bergman, Opuscula Physica et Chemica, pleraque ante seorsim edita, jam ab auctore collecta, revisa et aucta. Holmiae, Upsaliae Aboae, &c. Vol. I., 1779, II., 1780, III., 1786, IV., 1787, V., 1788, VI., 1790.

† Vol. III., p. 186.

"If x is greater than y there appears still no obstacle to prevent complete precipitation.

"But if x is less than y, so that only a part of B can be displaced, a portion of the dissolved precipitant must be sensibly thrown down, so as to act anew, or some other assistance must be given.

"II. Let $n > 1$ et $m < nm$

"With respect to phlogiston, this is the same result as in case I., but the obstacles are less."

(That is, if the acid for dissolving the precipitating metal is less than the acid which dissolves the metal to be precipitated, as in this case, the precipitating metal would not cease its action for want of acid.)

"III.) Let $n < 1$ then is $m > nm$. In this case B cannot be entirely thrown down, unless $nx = y$ or $nx > y$, because only n 100 of the precipitant A is dissolved."

(That is, if it requires more acid to dissolve the precipitating metal than the one in solution, then the metal in solution cannot be quite thrown down, unless it should be found that the amount of phlogiston in the precipitant is equal to the amount in the precipitate, or greater than it.)

Then, after recounting experiments, the first of which are made with a nitric acid solution, he says, p. 139;

"Therefore 135 parts of mercury have reduced completely into the metallic form by means of their phlogiston, 100 parts of silver which had been dissolved and calcined. This had united with four times its weight of mercury, and crystallized in an arborescent form.

"The amount of lead necessary for precipitating 100 lbs. of dissolved silver, amounts to 234 lbs. * * * * * * *

"C. 375 lbs. of shining plates of copper were put into a solution of silver, and were soon covered with a crystalline silver coating. When all the silver had fallen, the copper plates, when well cleaned, were found to have lost 31 lbs. The precipitated silver was found to amount to a cwt. (100 lbs.)

"In order to examine into the power of different solvents, we precipitated with copper a hundredweight of silver, which was dissolved in vitriolic acid, but there were only 30 lbs. of copper used. This, then, enables us, to some extent, to measure the great avidity with which nitric acid seizes on phlogiston, so much excelling the vitriolic acid."

The amount of each metal needed to precipitate 100 lbs. of silver, is given with the experiments, but to save room, I add a list.

"135	lbs.	Mercury dissolve	100 lbs. of Silver.
234	,,	Lead	ditto.
31	,,	Copper (with nitric acid)	ditto.
30	,,	——— (with vitriolic acid)	ditto.
29	,,	Iron (with vitriolic acid)	ditto.
88 ?	,,	Tin	ditto.
		Bismuth could scarcely be determined.	
64	,,	Nickel	ditto.
92	,,	Arsenic	ditto.
37	,,	Cobalt	ditto.
55	,,	Zinc	ditto.
83	,,	Antimony	ditto.
51	,,	Manganese	ditto.

Amounts of zinc used to precipitate 100 lbs. of metals.

217	lbs.	Zinc precipitated	100 lbs.	of pure Gold.
416	,,	,,	ditto	Platina.
44	,,	,,	ditto	Mercury.
26	,,	,,	ditto	Lead.
164	,,	,,	ditto	Copper.
68	,,	,,	ditto	Tin.
49	,,	,,	ditto	Bismuth.
70	,,	(the solution was difficult)	ditto	Antimony.

Scarcely any precipitation appears with Iron."

Then, at p. 150, there are certain corollaries, of which the following sentences suit best the subject in hand:—

COROLLARIES.

"A. That dephlogisticated metals unite with different acids in a variable manner (*i. e.*, that different amounts of metal unite to different acids). Thus, 100 parts of silver, dissolved in nitric acid, are reduced by 31 of copper, but if united to vitriolic acid, they want only 30 of copper. In the same way 100 parts of copper, in a vitriolic solution, are restored *to a metallic form* by 146 pounds of zinc, but in a nitric acid solution, 164 lbs. of zinc are wanted. Therefore nitric acid dephlogisticates the metals most, vitriolic acid less, and muriatic acid still less.

"B. Since we added the solutions in a saturated state, it is clear that the mutual quantities of phlogiston in the precipitate and the precipitant are in inverse proportion to the weights. Let the quantity of the phlogiston in a hundredweight of silver be 100, and the amount in a hundredweight of mercury will be 74, in lead 43, in copper 323, in iron 342, in tin 114, in bismuth 57, in nickel 156, in arsenic 109, in cobalt 270, in zinc 182, in antimony 120, in manganese 196. * * * *

"D. Let us see then how those principles before-mentioned may be applied. With respect to silver, which answers to *B*, there is no precipitant or *A*, which acts so as to make $n = 1$. If mercury or lead is used, then $n \mathrel{7} 1$, but if copper, iron, tin, bismuth, nickel, arsenic, cobalt, zinc, antimony, or manganese is used, the case is $n \angle 1$. In the zinc precipitates $n = 1$ is also wanting. Gold, platinum, iron, and antimony, make $n \mathrel{7} 1$, all the rest $n \angle 1$.

"Page 155. According to the experiments produced, the metals richest in phlogiston, are platina, then gold, iron, copper, cobalt, manganese, zinc, nickel, antimony, tin, arsenic, silver, mercury, bismuth, and lead, so that, in some order, it approaches nearer to the first metal. The relative numbers designating the amount found in each, are to be sought by other methods. A trial of each of the metals, so as to obtain the attractions sought for, would be a great labour, if done with sufficient care and sufficiently repeated, but if the work were divided it would be easier. If one would choose for examination mercury, another lead, a third copper, and so on, so as to see their relation with respect to the others, then we should have a series of experiments, which, if rightly compared, would not only disclose the various properties of each, worthy of observing, but would determine also the relative quantities. In this way, if the absolute value of only one were diligently sought out, all the rest would follow.

Vol. ii., p. 373. "The calces of metals have not that amount of phlogiston which is necessary to the metallic condition, but they are still found not entirely deprived of it.

"Metallic precipitates, when properly examined, reveal to us various mysteries."*

"In the following table 100 parts of reguline metal are in all cases understood to be dissolved:— †

100 parts of	Gold	with the	aerated mineral alkali gave	106	of dry precipitate.
,,		,,	caustic	110	,,
,,		,,	martial vitriol	100	,,
,,	Platina	,,	aerated mineral alkali	34	,,
,,		,,	caustic	36	,,
,,	Silver	,,	aerated mineral alkali	129	,,
,,		,,	caustic	112	,,
,,		,,	phlogist. (pruss. of pot.)	145	,,
,,		,,	saline	133	,,
,,		,,	vitriolated	134	,,
,,	Mercury	,,	aerated mineral alkali	110	,,
,,		,,	caustic	104	,,
,,		,,	vitriolated	119	,,
,,	Lead	,,	aerated mineral alkali	132	,,
,,		,,	caustic	116	,,
,,		,,	vitriolated	143	,,
,,	Copper	,,	aerated mineral alkali	194	,,
,,		,,	caustic	158	,,
,,		,,	phlogisticated	530	,,
,,	Iron	,,	aerated mineral alkali	225	,,
,,		,,	caustic	170	,,
,,		,,	phlogisticated	590	,,
,,	Tin	,,	aerated mineral alkali	131	,,
,,		,,	caustic	130	,,
,,		,,	phlogisticated	250	,,

* Page 390. † Vol. II., p. 390.

100 parts of	Bismuth	with the	aerated mineral alkali gave	130	of dry precipitate.
,,		,,	caustic	125	,,
,,		,,	phlogisticated	180	,,
,,		,,	pure water	113	,,
,,	Nickel	,,	aerated mineral alkali	135	,,
,,		,,	caustic	128	,,
,,		,,	phlogisticated	250	,,
,,	Arsenic	,,	phlogisticated	180	,,
,,	Cobalt	,,	aerated mineral alkali	160	,,
,,		,,	caustic	140	,,
,,		,,	phlogisticated	142	,,
,,	Zinc	,,	aerated mineral alkali	193	,,
,,		,,	caustic	161	,,
,,		,,	phlogisticated	495	,,
,,	Antimony		aerated mineral alkali	140	,,
,,		,,	caustic	138	,,
,,		,,	phlogisticated	138	,,
,,	Manganese		aerated mineral alkali	180	,,
,,		,,	caustic	168	,,
,,		,,	phlogisticated	150	,,

"Having compared the weights now produced, it is necessary, first, to inquire into the cause of such differences."

* * * * * Is it not then the matter of heat that is attached to the calx, and which is always united to the caustic alkali, for does it not excite heat when dissolved in the simple acids?

"This forms a triumphant foundation for assaying minerals and metals in the humid way, the mere weight of the precipitates being known. * * * * If the same mode of operating be used, the results of the experiments will be always the same. Let us say that a quantity of metal *a* in certain circumstances makes a precipitate of the weight of *b*; if the same method be used, it is obvious that *nb* may safely be allowed to correspond to *na* of the perfect metal, although in the fundamental experiment, the dissolved metal may not have been completely precipitated, or its weight may have been increased by foreign matter, still the same circumstances will produce always the same gain or loss, and the conclusion will remain unshaken. Let then the methods be exactly decided on, and no fallacy is to be feared." *

ON ELECTIVE ATTRACTIONS.

" Simple elective attractions.† Let *A* be the substance to which *a*, *b*, *c*, &c., are drawn; further, let *A* be added to *c* to saturation, this we shall call *Ac*, when again *b* is added, let the union take place to the exclusion of *c*; then *A* is said to attract *b* more than *c*, or *b* has a stronger elective attraction than *c;* then *Ab*, when *a* is added, gives up *b*, and *a* is united instead, then it is understood that *a* excels *b* in attractive force, and the order of the efficacy of the attraction forms a series *a*, *b*, *c*. What we here call attractions, others call affinity; we use either term promiscuously in future, although the latter being more a metaphorical expression does not appear so suitable in physics.

* * * * Page 318. " It has not escaped me, that some chemists have considered, as entirely without foundation, the doctrine which asserts that neutral or middle salts can receive a distinct excess of acid. That this sometimes takes place the experiments to be related most clearly shew, although naturally it *(the excess)* adheres with far less tenacity than that portion which is necessary to effect saturation.

* * * * Page 325. " From all that has been brought forward, I consider it clear, not only that the doctrine of a decided superfluity of either ingredient is not absurd, but that in reality this result is found in many cases. Certainly this superfluity attaches itself much more loosely than the portion necessary for saturation, so that frequently it is easily driven off, but this in no way causes it to be less real."

Here we find Bergman endeavouring to obtain the amount of oxygen in metallic oxides, or phlogiston in metals. He finds that the amount in equal quantities of metals is not the same. This could only be the case if the atomic weights of all metals were the same.

* Page 396. † Vol. III., page 294.

In agreement with this, he finds that the quantity of acid necessary for dissolving certain weights of metals, differs with each metal, and the amount which one metal precipitates from the solution, differs with each metal. This was promising fair for discovery; and in the first table we have the amount of various metals needed to precipitate 100 of silver, in fact, a table of atomic weights, if he could have seen it, although imperfection in experiment rendered it difficult, and the law seemed very intricate.

He drew the conclusion that some acids dissolve metals with more oxygen in their oxides than others, when he says, 100 of silver are reduced by 31 of copper in nitric, and 30 of copper in sulphuric acid. This helped to lead him wrong.

He seems to have most naturally thought that it would be needful to find the relation of the oxygen in a metallic oxide to that in every other, and was naturally surprised at the great labour needed. We know that this would be a most complicated relationship, and that the oxygen is constantly changing its per centage relation in every compound, to such an extent, that it would be impossible to follow it without constant recurrence to its atomic weight. We may look on this inquiry of Bergman as a search, acute although unsuccessful after that last step in simplicity.

He gives a valuable discovery in the establishment of the permanence of the amount of oxygen in precipitated oxides, the very foundation of analysis, and an important step towards the knowledge of permanence of constitution in all substances whatever. That the numbers need correction, need hardly be remarked.

At the same time it seems to be beyond doubt that he did not grasp with great clearness the doctrine of permanent constitution, or he would scarcely have made these remarks on neutral salts receiving a distinct excess of acid. Any indefinite amount added, becomes a mixture only.

He extended the tables of attraction to a great length,

calling them elective attractions, preferring attraction to affinity. His tables are 59 in number, the first portion giving the wet way, and the second the dry. These were given in the old symbols, and have certainly a most formidable and unattractive appearance, in his original work. They have, however, been published in England, at an early period, in the form below.

SINGLE ELECTIVE ATTRACTIONS.

IN THE MOIST WAY.

SULPHURIC ACID.	NITRIC ACID.	MURIATIC ACID.
Potash	Potash	Potash
Soda	Soda	Soda
Baryta	Baryta	Baryta
Lime	Lime	Lime
Magnesia	Magnesia	Magnesia
Ammonia	Ammonia	Ammonia
Alumina	Alumina	Alumina
Oxide of Zinc	Oxide of Zinc	Oxide of Zinc
,, Iron	,, Iron	,, Iron
,, Manganese	,, Manganese	,, Manganese
,, Cobalt	,, Cobalt	,, Cobalt
,, Nickel	,, Nickel	,, Nickel
,, Lead	,, Lead	,, Lead
,, Tin	,, Tin	,, Tin
,, Copper	,, Copper	,, Copper
,, Bismuth	,, Bismuth	,, Bismuth
,, Antimony	,, Antimony	,, Antimony
,, Arsenic	,, Arsenic	,, Arsenic
,, Mercury	,, Mercury	,, Mercury
,, Silver	,, Silver	,, Silver
,, Gold	,, Gold	,, Gold
,, Platina	,, Platina	,, Platina
,, Water	,, Water	,, Water
,, Alcohol	,, Alcohol	,, Alcohol
,, Phlogiston	,, Phlogiston	,, Phlogiston

IN THE DRY WAY.

SULPHURIC ACID.	NITRIC ACID.	MURIATIC ACID.
Phlogiston	Phlogiston	Phlogiston
Potash	Baryta	Baryta

SULPHURIC ACID.	NITRIC ACID.	MURIATIC ACID.
Soda	Potash	Potash
Baryta	Soda	Soda
Lime	Lime	Lime
Magnesia	Magnesia	Magnesia
Metallic Oxides	Metallic Oxides	Metallic Oxides
Ammonia	Ammonia	Ammonia
Alumina	Alumina	Alumina

He gave also two tables of his famous double elective attraction, or compound attraction. The examples given are numerous, and would take too much room. The form is exactly the same as given below as Elliot's, no numbers being used.

Elliot published Bergman's tables, with the addition of figures, to show the relative force which one bore to another. He says, "suppose that (see Encycl. Method. Dict. de Chymie, vol. i., p. 552) potash and sulphuric acid attract each other with the force of 9; that oxide of silver and nitric acid attract each other with the force of 2; that the affinity of nitric acid, with potash, is 8, and that of sulphuric acid, with oxide of silver, 4. As 8 + 4 is greater than 9 + 2, decomposition takes place, and two new compounds are formed, nitrate of potash and sulphate of silver."

He then made the symbols so:—

		Nitre of Potash		
	Potash	8	Nitric Acid	
Vitriol of Potash	9		2	Nitre of Silver
	Vitriolic Acid	4	Calx of Silver	
		Vitriol of Silver		

G. Morveau continued this *schema* or *symboles*, finding new numbers, and he has put into a short table his results. This is a more definite way of showing the relation of bodies to each other than we have yet seen.

*Table of the numerical expression of the affinities of five Acids and seven Bases, according to the constant relations indicated by the most familiar observations.**

	VITRIOLIC ACID.	NITRIC ACID.	MURIATIC ACID.	ACETIC ACID.	CARBONIC ACID.
Baryta	65	62	36	29	14
Potash	62	58	32	26	9
Soda	58	50	28	25	8
Lime	54	44	20	19	12
Ammonia	46	38	14	20	4
Magnesia	50	40	16	17	6
Alumina	40	36	10	15	2

At the same time he says,† relating to the figures, " the numbers which I have employed have no certain basis, but because they agree with a sufficient number of the most familiar observations, they may be used without inconvenience, until we recognise the necessity of changing them, so as to make them agree with other results."

Fourcroy gave numbers also on similar principles, but Morveau objects to them as being so small that it was not easy to find intermediate ones, whilst he objects to Kirwan's numbers which gave the weight of the base as the amount of the affinity, because this did not agree with results.

In these schemes of double decomposition there seems to be a tacit agreement, that the acid which saturated one base, would saturate the second.

Kirwan experimented very much in the direction which Bergman had followed. He is another of those who nearly discovered the atomic theory, who laboured in a legitimate direction, but whose discoveries and theories on the subject are merged in the higher and simpler law.

A brief extract will give his results.

" The discovery of the quantity of real acid in each of the mineral acid liquors, and the proportion of real acid taken up by a given quantity of each basis at the point of saturation, led me, unexpectedly, to what seems to be the true method of

* Dict. de Chymie, Vol. I., p. 558. † Page 557.

investigating the quantity of attraction which each acid bears to the several bases to which it is capable of uniting; for it was impossible not to perceive, first, that the quantity of real acid, necessary to saturate a given quantity of basis, is inversely as the affinity of each basis to each acid. 2ndly. That the quantity of each basis, requisite to saturate a given quantity of each acid, is directly as the affinity of such acid to each basis. Thus 100 grains of each of the acids require for their saturation a greater quantity of fixed alkali than of calcareous earths, more of this earth than of volatile alkali, more of this alkali than magnesia, and more of magnesia than of earth or alum, as may be seen in the following table.

Quantity of Basis taken up by 100 *grs. of each of the Mineral Acids.*

	Veget. fixed Alkali.	Mineral Alkali.	Calcareous Earth.	Volatile Alkali.	Magnesia.	Earth of Alum.
	grs.	grs.	grs.	grs.	grs.	grs.
Vitriolic Acid...	215	165	110	90	80	75
Nitrous Acid ...	215	165	96	87	75	65
Marine Acid....	215	158	89	79	71	55

"As these numbers agree with what common experience teaches us concerning the affinity of these acids with their respective bases, they may be considered as adequate expressions of the quantity of that affinity, and I shall in future use them as such. Thus the affinity of vitriolic acid to fixed vegetable alkali, that is, the force with which they unite, or tend to unite, to each other, is to the affinity with which that same acid unites to calcareous earth, as 215 grs. to 110; and to that which the nitrous acid bears to calcareous earth as 215 grs. to 96," &c.* He adds a similar table of metals and acids. Kirwan gives here what would lead to the atomic weights of the bodies had he known the law which appears to have been first published by Richter; one obtained the atomic weights as the measure of affinities, the other reciprocal

* Philosophical Transactions Abridged, Vol. XV., p. 335-6, year 1783. This was read, I believe, in 1782.

affinities, but neither knew the other's results, and both were lost sight of. The one (Ritcher) did not know that he had got close upon a universal law, the other (Kirwan) did not know that he had got the mode of expressing that universal law, but used it for what it was little worth, an expression of affinity.

We now come to Wenzel, one of those men whose names have been brought forward as a much neglected philosopher, and to whom almost every writer on the history of science, who has had occasion to mention him in later years, has been anxious to award the due honour. We see his book constantly quoted. Some writers give us his words, others give us what appears such a clear explanation of his ideas that we feel no more to be wanting. I had been long anxious to obtain his works, but after advertising in Germany, and inquiring in several towns and large libraries in this country, as well as in France and Germany, I did not obtain the volume, and proceeded without it. I afterwards found that a duplicate copy existed at the Munich Royal Library, and was fortunate enough to obtain it, duplicate copies being generally disposed of. Having read it carefully over, I found no such passages as are imputed to him; and, therefore, read it still more carefully again, and then a third time, but they did not exist. Having written to two eminent historians of science for an explanation, I find that neither had seen the volume; but one of them informed me that the mistake had been rectified in a supplement to the "Handwoerterbuch der Chemie u. Physik."*

The reciprocal saturation which results when two salts decompose each other, is the discovery, the honour of which has long been given to Wenzel. It is a curious fact that not only does he not see this, but he sees and explains the con-

* It is by Dr. J. S. C. Schweigger, and has been since published as a pamphlet (Ueber die Stoechiometrische Reihen im Sinne Richter's), &c., Halle, 1853.

trary, as he shews us that in double decomposition something always remains unsaturated; but generally very little remains. One is sorry that being so near a law, he had not the slightest conception of it. The most important part of his work, as far as our purpose is concerned, seems to me to be contained in the following sentences. The title of the work is "The Doctrine of the Affinity of Bodies."* I shall not give the original, although scarce, as the work, from the fact above stated, has lost its great importance.

In the Preface, he says, "at first my only intention was to make for my own use a treatise which should contain the order of the ascertained affinities and the circumstances under which they acted, lest I should not be able to remember them. But it occurred to me that others might find it useful also, if it were more worked out. For this end I endeavoured to explain the cause and the law of affinity on a good foundation, and the circumstances under which the bodies combine as well as the true relation of their weights towards each other.

Page 4. "It is of itself clear that any combination of bodies must have a constant unchangeable proportion, which can neither be greater nor smaller without some cause acting externally, because, otherwise, nothing certain could be decided on by comparing them. It therefore necessarily follows, that every possible combination of two bodies stands in the most exact relationship with every other, and this relation expresses the degree of combination.

Page 9. "These smallest particles of each body have at all times, in a natural state, a determinate figure; but the whole mass of the body takes a form according as chance or art gives it, without causing any change in the smallest particles, just as the tender fibres or tubes in a piece of wood remain always the same, although the whole piece may be in the shape of a ball or a cube."

* Carl Friedrich Wenzel, Lehre von der Verwandschaft der Koerper. Dresden, 1777.

Page 10. "I examine the natural structure of some metals, I see certainly nothing more than that they are hard solidly-united heavy bodies, which become liquid in the fire at different degrees of heat, and lose their former connectedness (or cohesion), and without being heavier take up a greater or less space than before. This is enough to enable us to conclude that the figure of the smallest particles of metals is changed by the fire, and that the fluid condition of the whole mass, and its altered specific gravity, are the necessary consequence of this alteration of figure. For when the mass of a body without change of weight takes up a greater or less space, it is certain that it can take place under no circumstances except a change of figure in the smallest portions of the bodies. A thousand small cubes may be put into a smaller space than the same number of spheres of the same mass and weight, and the heap made by the spheres is not so great as if they were converted into stars, and so on. When the specific gravity is altered, no matter by what means, the figure and situation of the smallest parts can no longer remain the same."

Page 20. "Besides change of figure, I know no sufficient reason for all that has been said; for if we completely banished the figure, and viewed the properties of the body as something substantial in matter, I know not how we could explain without contradiction the every-day experience; or we must, as Snellius with refraction, explain it by the will of God, which settles the matter at once; but if my understanding is to lay hold of the method by which anything acts, this explanation will not be satisfactory.

Page 28. "But we have remarked that any combination of bodies, on account of the figure of their parts, depends on static laws, and there it is proved that the motion of a weight is so much the slower the smaller the force is, in comparison with it. Let us apply this to the present case, and bodies will appear to us as so many weights, and their common solvent as a force, which acts more slowly or more rapidly on one or the other. It follows, then, that the more rapidly a common solvent unites with a body, the greater must be its degree of combination, and we obtain therefore this law.

" *The affinity of bodies with a common solvent is in the inverse ratio of the time taken to dissolve.*

Page 31. " We have now a universal law, according to which the affinity of bodies or their rank in the series is decided, and we obtain at once this important advantage, that we not only know that the union of a common solvent is greater or less with any body, but also how much greater or less it is, because the difference of the time of solution shews the difference of the combination. Therefore amongst a number of bodies, the combination of one with a common solvent may be considered as a quantity which may be expressed by a fixed number, if we take the smallest in such a series as *unity*; and by this means we are able to give a correct explanation of all phenomena.

Page 46. " This important question, then, remains, why a solvent, when it is only moderately diluted, does not in the least attack certain metals, but as soon as another metal is dissolved in it, with which it naturally has a less affinity, a ready solution of the first takes place. Page 47. Because here the powers meet which assist each other.

Page 72. " The circumstances under which this metal (iron) is dissolved by vitriolic acid are these, that the acid must not be strong. When both unite, iron vitriol is formed, which loses the most of its acid in the fire, as well as by frequent solution in water. A small bored cylinder of Styrian steel of 102 grains was put into half an ounce of the spirit of vitriol, diluted with an equal quantity of water, exactly as with the zinc experiments; there remained $46\frac{3}{4}$ grains of steel, and $55\frac{1}{4}$ grains were dissolved in the half ounce of the spirit of vitriol.

"Therefore the relation of the hardest steel to the strongest vitriol is 175 : 240.

"*Application of the doctrine of affinity of bodies.** "This will best be shewn by examples.

"Is it possible by Beguin's spirit of sulphur (a sulphide of ammonium chiefly) to decompose the *luna cornua*, or to separate the muriatic acid entirely without loss?"

"To settle this question we require only the following experiments. Muriatic acid has a smaller degree of affinity with silver than with the volatile salt. Sulphur, on the other hand, unites with silver in preference to the volatile salt. The silver is not separated from the muriatic acid by the volatile salt, on account of accidental circumstances, but this separation follows the moment any other body unites with the silver, if it has not the property of dissolving the silver. But sulphur is just such a body, and is, therefore, fitted for the purpose. If, then, the spirit of sulphur of Beguin is poured on finely powdered luna cornua, it is easily seen that the muriatic acid in the luna cornua must unite with the volatile salt in the spirit of sulphur, and the sulphur will unite with the silver. The new products that are formed by this separation, can consequently be nothing more than common sal ammoniac and sulphuretted silver.

Page 452. "Another similar question arises by which the proportions of the mixture must be considered. How much cinnabar must be mixed with the *luna cornua*, so as completely to separate the silver?

"The possibility of this decomposition may be shewn in the same way as in the first case. If no particular experiment is made, it depends on the comparison only of the following propositions. Half an ounce of *luna cornua* contains $180\frac{9}{16}$ grains of fine silver. Half an ounce of fine silver takes up $35\frac{1}{2}$ grains of sulphur. We may then calculate that for $180\frac{9}{16}$ grains of fine silver, $26\frac{3}{4}$ grains of sulphur are required. We know besides,

* Page 450.

that cinnabar contains sulphur in the proportion of 65 to 240 of quicksilver, or 65 grains of sulphur united with 240 of quicksilver, are to be met with in 305 grains of cinnabar, therefore $26\frac{3}{4}$ grains of sulphur are contained in $125\frac{1}{2}$ of cinnabar. This quantity of cinnabar, as regards its sulphur, will be sufficient for the decomposition of half an ounce of *luna cornua*.

"But we must inquire if $125\frac{1}{2}$ grains of cinnabar contain as much quicksilver as will be sufficient to take in the muriatic acid which is saturated with the silver. Half an ounce of luna cornua contains $53\frac{7}{16}$ grains of muriatic acid of greatest concentration. In half an ounce of the caustic sublimate there are $58\frac{1}{3}$ grains of the strongest acid, which is saturated with 174 grains of quicksilver. From this proportion it is found that $53\frac{7}{16}$ grains of the strongest muriatic acid are required for $159\frac{2}{5}$ grains quicksilver. Now as there are in cinnabar 240 grains of quicksilver united with 65 grains of sulphur, $159\frac{2}{5}$ grains of quicksilver require $43\frac{1}{8}$ grains of sulphur. Both together give nearly $202\frac{1}{2}$ grains of cinnabar. Consequently, from $125\frac{1}{2}$ grains of cinnabar, all the muriatic acid found in the luna cornua is not separated. We see from this that the muriatic acid of the lunar caustic rises in sublimation with the quicksilver out of the $202\frac{1}{2}$ grains of cinnabar as a caustic sublimate, whilst the silver remains united only with so much sulphur as it found in $125\frac{1}{2}$ grains of cinnabar."

His *smallest parts* of bodies are not atoms, but molecules rather, or particles, as they change their form.

He has made a theory of affinity, and attempted to represent the force by a number. To attempt to give the numerical or dynamical ratio of every body to each other was an object of the very highest kind, and we must look on him as one of those less fortunate men who, when search was required in every direction, has had the misfortune to have the wrong one assigned to him. He searched in the direction of time, and obtained a manifest fallacy; as bodies are constituted abstractly he might be correct, but his theory cannot be introduced into science at present, and in the way he introduced it, it was entirely a mistake. But he has done great service in early times in seeking for the distinct constitution of bodies, and in asserting the constancy of combination; whilst he obtained numbers representing the constant relation of bodies to each other, he failed to see that they would be reciprocal. This failure at once removes him from the great discoverers, and places him among those honourable and valuable labourers in science whose names are read with respect by students, but who cannot be recognised by mankind generally, because the capacities of our minds are too small to retain more than the lives of a few of the most eminent.

The doctrine of reciprocal proportion must be taken from him, and he can now no longer hold a place in the history of the atomic theory other than as the author of an intelligent attempt which has entirely failed.

I feel sorry to leave him in this state, and a few kind words will do little good. I believe he would have preferred the truth; the honour he received was not required by him; the discovery was not claimed by him; he died in 1793, before it was known to be worth making. In his works he appears an honest, earnest man.

CHAPTER VIII.

DR. BRYAN HIGGINS AND WILLIAM HIGGINS.

Dr. Bryan Higgins was the earliest chemist who seems to have dealt with the elements of matter, in a chemical, as well as physical sense, with an attempt to obtain information as to the primary elements and their atomic constitution, although he gives no certain results. I find at the laboratory of Owens College, among the books forming part of the library of the late Dr. William Henry, a little pamphlet, proposing a course of lectures, to be commenced on the 13th November, 1775.

The proposals are said to have been formerly published; it is not said how long before. I shall give considerable extracts.

Page 1. "Doctor Higgins, of Greek-street, Soho, encouraged by the literary noblemen, and gentlemen who have subscribed to his courses of philosophic and practical chemistry, addresses the following proposals to the patrons of natural philosophy and useful arts.

"That fifty philosophic and literary gentlemen do concur in promoting experimental inquiries into the elements of matter and laws of nature, and such other subjects as are most important in natural philosophy, chemistry, and arts.

Page 3. "That in these discourses he shall introduce the natural phenomena, the illustrative observations and experiments of philosophers, chemists, and artists; and particularly his notions and experiments concerning the primary elements and the properties of matter.

Page 9. "Introductory discourse on matter in general, called gross matter; on the varieties and distinctions of gross matter; on the primary elements of matter.

"Observations on the experiments and philosophy exhibited in the foregoing course of chemistry, and other experiments which demonstrate the existence of seven primary distinct elements of matter, viz.;

Earth,	Air,
Water,	Phlogiston,
Alkali,	Light.
Acid,	

"Experiments, observations, and arguments, persuading that each primary element consists of atoms homogeneal; that these atoms are impenetrable, immutable in figure, inconvertible, and that, in the ordinary course of nature, they are not annihilated, nor newly created.

"Observations and experiments, persuading that the atoms of each element are globular, or nearly so, and that the spiral, spicular and other figures ascribed to these atoms, are fictions unnecessary, and are inconsistent with the uniformity and simplicity of nature, and repugnant to experience. * * *

"Experiments and observations showing that the possible and known unions of the foregoing elements, and that the possible and known proportions in which the unions of the foregoing elements may take place, are more numerous than the bodies distinguished by philosophers and naturalists; persuading that all known bodies are really composed of one or more of the foregoing elements; and that all bodies must be admitted to consist of these only until other elementary matter is found necessary for the explication of the natural phenomena, and is demonstrated to exist.

"A classical arrangement on the table of bodies composed of two or three primary elements, which bodies, in various chemical processes, not being decomposed, we call chemical elements, or the elements of the chemists.

"A like classical arrangement of bodies composed of two chemical elements.

"A like classical arrangement of bodies and natural substances composed of many chemical elements.

OPINION,

"1. That the homogeneal atoms of five elements repel reciprocally.

"2. That the homogeneal atoms of two elements attract reciprocally.

"3. That the dissimilar atoms of five elements attract reciprocally.

"4. That the dissimilar atoms of two elements repel reciprocally.

"That the attraction subsisting between elementary atoms is more forcible in one direction or axis of each atom than in any other direction, and that there is a polarity in all matter whatever.

"6. That there is but one species of attraction operating with great force between the similar or dissimilar atoms of certain elements, and with less force between those of other elements, in gradations; but in all affected by distance and polarity.

"7. That the attraction of bodies enumerated as distinct properties of matter or laws of nature, are nothing more than the sums of the attraction of their elementary atoms, or these forces concentrated in a certain degree by the pressure of repellent atoms, or these forces exerted to the greatest advantage in bodies whose primary elementary attractions are strongest, and whose primary elementary atoms are also arranged in polar order.

"8. That specific gravity is not as the quantity of matter in a given space, but as the quality of the matter, or the sum of its elementary attractions; consequently, that light bodies are not necessarily more porous than the heaviest."

Page 14. "Observations and experiments, showing the grounds on which we ought for a while to admit the following distinctions of earths, viz:

"Seven earths, capable of forming ductile metals.

"Seven earths, capable of forming metals not ductile.

"Seven earths, incapable of forming metal.

Question 1. Is there but one earthy element, which, in various modes of aggregation, or in indissoluble combination with other elementary matter, forms twenty-one earthy bodies? or, Question 2, Are there three times seven, or seven times seven earthy elements?

Page 15. "Experimental and geometrical estimation of the force of this attraction in fortuitous arrangement of the atoms, and of the force of this attraction in the polar arrangement of the atoms.

On attraction he says, page 23, "That no element doth saturate, nor can saturate, the like element; that no element, whose atoms attract each other, can saturate any other element whose atoms attract each other; that a repellent element doth saturate non-repellent elements, and *vice versa;* that repellent elements do saturate reciprocally; and that attraction and repulsion, operating adversely, are the cause of saturation; and saturation is not a distinct or primary law of nature, but an effect."

We find here no ideas given of definite compounds, except so far as the ordinary idea of saturation is concerned. Had Dr. Higgins any theory resembling the present atomic theory, he would certainly not have expressed himself so darkly on the subject of saturation.

We even find that he is not quite freed from the *prima materia*, although he restricts it to the matter of certain classes, such as the matter of earth forming many earths, the acid matter forming many acids. He is, therefore, to be viewed as one bordering on the transmutation theory, not freed from mystic ideas, but grappling with the subject so energetically, that in some directions he almost sees his way into another region of theory.

I shall quote the most important sentences in the "Experiments and observations"* of Dr. Higgins. In distilling acetate of lead, he found a certain amount of what he calls acid matter in the fixable air, which he considers, as before mentioned, to be a peculiar principle. It combines with the empyreal air (oxygen) of the litharge. Not accounting for the whole amount by the measurements he made, he inferred "that when the acid matter of acetous acid is employed in excessive quantity to form fixable air with the empyreal air of litharge, the fixable air may consist of a little more than one part of the acid matter, combined with two of the empyreal air. By a more accurate estimate of the fixable air taken at 85 grains, it is most probable that the proportions would be found to be accurately two to one, provided fixable air, like other acids, may not subsist with various proportions of the empyreal air." †

He mentions the definite proportion, two to one, because he obtains the figures approximatively so, but he not only fails to elevate it into a principle, but speaks of various proportions as probable in the case in hand, and as usual in other cases. This seems equal to saying distinctly that he recognised no such principle. But does the term "various proportions" allude to fixed numbers, such as two or three? There is no reason to suppose this, he uses the word in the ordinary sense it was then used, no other sense had been given to it, and any other sense in this place is impossible.

When firing with oxygen the inflammable gases from acetate of lime, he says, in reference to some experiments unnecessary to be detailed, "by other experiments and the same kind of estimation, the empyreal air appeared to constitute more than two-thirds of the fixable air, and in some it seemed to be accurately two-thirds; but after all I continued

* Experiments and observations relating to acetous acid, fixable air, dense inflammable air, oils and fuel. The matter of fire and light, metallic reduction, combustion, fermentation, putrefaction, respiration, and other subjects of chemical philosophy. By Bryan Higgins, M.D. London, 1786.

† Page 232.

to suspect that the proportions of the principles are not the same in every specimen of the elastic fluid which we consider as fixable air. When the dense inflammable air first expelled from charcoal was used, the result was nearly the same. And from the whole, I conclude, that when as much inflammable air of this kind is employed, as can be converted in the explosion, by the empyreal air, the fixable air consists of one part by weight of the acid matter of acetous acid or charcoal, and nearly, or accurately, two parts of empyreal air; and almost one-fifth of this kind of inflammable air is phlogiston, and the remainder mere acid basis of fixable air, and of acetous acid, charcoal, oils, spirits, and of all substances that yield acetous acid or dense inflammable air abundantly."*

The same remarks apply to this. The "nearly" two parts are just as probable, in his mind, as the "accurately" two parts, which would not have been the case had he found any defining law to express on the subject.

Then he says, further on,† "It seems, therefore, that the proportions in which the acid and phlogistic matter are combined, in different specimens of inflammable air expelled from vegetable substances, do vary considerably," &c. This refers to his tables of results.

He believes that the particles of the different gases unite to form molecules of compound gases, page 317: "I consider the specific gravity as a safe guide in our investigation of these affinities and of their order, in regard only to the elastic fluids which seem to consist of no more than one kind of gravitating matter engaged in the repellent atmospheres; and of fixed air, dense inflammable air, acid air, the phlogistic alkaline air, and others, I would observe, that the atmospheres include molecules, instead of solitary ultimate parts; for, without this chemical union of heterogeneal parts, and the formation of molecules, an elastic fluid of the kind that I now speak of could not differ, as it does, from either kind of matter

* Page 291. † Page 294.

of which it is composed. From this consideration of the attractive forces which tend to form molecules, and of the atmospheres which, in compound elastic fluids, encompass the molecules, but not the ultimate parts severally, we derive an easy explanation of the phenomenon so often noted in the preceding pages—I mean, the conversion of a substance, not into air, but into two or three different elastic fluids, by mere ignition."

Dr. Higgins thinks of atoms, of simple particles, and even speaks of gases uniting, in some cases, in nearly, if not accurately, a fixed proportion, and yet he sees no law. He does not carry his idea far enough. If the molecules are formed by the union of two particles the proportions must of necessity be fixed, or if any number of particles unite, the proportions are fixed, unless the molecules are to be supposed of different constitutions. In this last case, they would constitute a mixture of different gases. But it was not known at this period that all bodies had a fixed constitution, otherwise one would have supposed that Dalton's laws would have been readily arrived at by Dr. Higgins. On the other hand, his theory was not clear, or he would have been led by it to decide on the necessity of fixed constitution as a result. But we obtain no results affecting chemical philosophy. And yet amongst other questions which he says "we shall find no difficulty in answering," is, "Why does any excessive quantity of empyreal or of inflammable air, beyond the determinate proportions in which their gross parts can combine, remain elastic and unaltered, or not altered in any considerable part of it, after the combustion?" *

The nearest answer he gives as to the proportions is the following:—"The matter of fire limits the quantity in which aeriform fluids, and bodies containing it, can combine chemically." † We may conclude then that nothing in the gases themselves determined the proportion, but the cause was in "distinct atmosphere of fiery matter."

* Page 330. † Page 307.

In reading the first part of this question we suppose he must live as a discoverer; in reading the second part, we find him timidly committing suicide. He wavers between his theory and his experiments. If determinate proportion be a law, he might have supposed it rigid, like other laws of nature; but he leaves a portion of the residue altered, so that, after all, we only receive from him indeterminate, or nearly determinate proportions.

The experiment to which the question refers occurs in p. 296. He draws no valuable conclusion from it. He says also, p. 299, "In the combustion of charcoal with empyreal air, the expenditure of the latter, in fixable air and water, was always found to be more than thrice the weight of the charcoal. I could now easily ascertain the proportion of these, and even the quantity of the acid and phlogistic matter in this and other bodies; but as my present purposes are answered by approximation, I think it unnecessary to detain the reader any longer on this subject."

His principal object was to explain the nature of fire, which he considers as subject to the laws of gravitation, and to be the cause of the aeriform state of bodies.

The only law in which he introduces numbers is the fourth of his "primary notions of the matter of fire." "The changes of repellent matter, by which attractive and gravitating particles form elastic fluids, are distinct atmospheres of fiery matter, in which the densities are reciprocally as the distances from the central particles, in a duplicate or higher ratio." *

His writings are mostly in the first stage of thought before opinion is formed. Commentators on such persons are obliged to extend the ideas a little in order to make them clear, and so the original writer gets credit for more than he had ever done. Such writers are of great value when they lead towards discoveries; but we are apt to give them the entire honour when

* Page 306.

they deserve only a part. As far as our subject is concerned Dr. Higgins has small claims. His opinions on atoms might have been held by the ancients: whilst standing on their shoulders, it would have required much less sagacity to discover than was needed for them. He speaks of the sums of the forces of atoms measuring the attraction of matter, but does not suppose that if matter be atomic, the number of atoms might also, in this way, be got comparatively.

We might say, however, that Dr. Higgins began the study of atomic chemistry, properly so called, although he made little advance in it.

The next person worthy of note is of the same name and family. The following extracts contain what is of greatest importance in connection with our subject. The title page of the volume quoted from is, "A Comparative View of the Phlogistic and Antiphlogistic Theories, with Inductions, &c., by William Higgins, of Pembroke College, Oxford. The Second Edition.

"Est quoddam prodire tenus, si non datur ultra.

"London: Printed for J. Murray, No. 32, Fleet Street. 1791." The first edition was in 1789.

Page 14. "It is generally allowed, and justly, that nitrous air consists of dephlogisticated air and phlogistic, in the proportion of two of the former to one of the latter. The supposition of its containing phlogiston, I hope, will hereafter appear to be erroneous; therefore, every ultimate particle of phlogisticated air must be united to two of dephlogisticated air; and these molicules, combined with fire, constitute nitrous air. Now, if every (one) of these molicules were surrounded with an atmosphere of fire equal in size only to those of dephlogisticated air, 100 cubic inches of nitrous air should weigh 98,535 grains; whereas, according to Kirwan, they weigh but 37 grains. Hence, we may justly conclude that the gravitating particles of nitrous air are thrice the distance from each other that the ultimate particles of dephlogisticated are in the same temperature, and, of course, their atmospheres of fire must be in size proportionable; or else some other repelling fluid must interpose. The size of the repelling atmospheres of nitrous air thus considered, and likewise the weaker attraction of the molicules of this air to dephlogisticated air than that of the ultimate particles of phlogistic in their simple state, it is surprising to me, with how much more facility the former unites to dephlogisticated air than the latter."

After speaking of the combustion of sulphur, he says, p. 35, "A good many more facts might be urged on this subject; but, in my opinion, enough has been adduced to convince an impartial reader that all the phenomema above recited are only explicable by entirely leaving out phlogiston, and supposing sulphur to be a simple subtance, whose ultimate particles attract dephlogisticated air with forces inherent in themselves, independent of phlogiston or concrete inflammable air, as an alkali does an acid, or gold and tin mercury; and likewise supposing the combustion of sulphur to be as simple a process as that of light inflammable air; that is, that there is no dephlogistication, or formation of water, during the union of the oxygenous principle to sulphur, as containing not a particle of light inflammable air in its constitution. I have often combined sulphur rendered perfectly dry, and dephlogisticated air likewise, deprived of its water by fused marine selenite in large proportion over mercury, and could never observe that water was produced. Indeed it may be said that the volatile sulphurous acid, which is always the result of this process, may redissolve it; but this is not very likely, when a small portion of water will deprive it of its elasticity."

"According to Mr. Kirwan, 100 grains of sulphur require 143 grains of dephlogisticated air to convert them into volatile vitriolic acid; but they require much more in order to become perfect vitriolic acid. Highly concentrated vitriolic acid contains two parts of dephlogisticated air, and one of sulphur, exclusive of water. One hundred and forty-three grains of dephlogisticated air contain 41 of water, for lime will abstract 26 grains from it, and the remainder cannot be separated from it in its aerial state; therefore 100 grains of sulphur, making an allowance for water, require 100 or 102 of the real gravitating matter of dephlogisticated air to form volatile vitriolic acid; and as volatile vitriolic acid is very little short of double the specific gravity of dephlogisticated air, we may conclude that the ultimate particles of sulphur and dephlogisticated air contain equal quantities of solid matter; for dephlogisticated air suffers no considerable contraction by uniting to sulphur in the proportion merely necessary for the formation of volatile vitriolic acid. Hence we may conclude, that, in volatile vitriolic acid, a single ultimate particle of sulphur is intimately united only to a single particle of dephlogisticated air; and that, in perfect vitriolic acid, every single particle of sulphur is united to two of dephlogisticated air, being the quantity necessary to saturation."

"As two cubic inches of light inflammable air require but one of dephlogisticated air to condense them, we must suppose that they contain equal number of divisions, and that the difference of their specific gravity depends chiefly on the size of their ultimate particles; or we must suppose *that the ultimate particles of light inflammable air require two or three, or more, of dephlogisticated air to saturate them. If this latter were the case, we might produce water in an intermediate state, as well as the vitriolic or the nitrous acid, which appears to be impossible; for in whatever proportion we mix our airs, or under whatsoever circumstance we combine them, the result is invariably the same.* This likewise may be observed with respect to the decomposition of matter. Hence we may justly conclude, that water is composed of molicules formed by the union of a single ultimate particle of dephlogisticated air to an ultimate particle of light inflammable air, and that they are incapable of uniting to a third particle of either of their constituent principles. The above notions of water and vitriolic acid being strictly kept in view, let us now proceed to inquire into the nature of sulphur and vitriolic acid, and their various effects on different bodies in the antiphlogistic doctrine. It has been already observed that metals attract dephlogisticated air with greater force than sulphur, and that sulphur attracts it with greater force than light inflammable air. It has likewise been observed, that vitriolic acid and water, mixed in a certain proportion, will calcine metals with greater facility than concentrated vitriolic acid, and that water will have very little effect on metals in a common temperature. These facts, though they may appear contradictory in themselves when slightly considered, may be accounted for on the following principles, and are, in my opinion, inexplicable by any other means whatever."

"Let us suppose iron or zinc to attract dephlogisticated air with the force of 7, sulphur to attract it with the force of 6 7-8ths, and light inflammable air with the force of 6 5-9ths. Let us again suppose these to be the utmost forces that can subsist between particle and particle. That is to say, in water dephlogisticated air is retained with the above force, and likewise in volatile vitriolic acid, with the force already mentioned. It is unnecessary to introduce here the aggregate attraction which frequently preserves a neutrality between bodies, as, for instance, between water and zinc, or water and iron. Stating the attractive forces in the above proportion, which I am led to believe is just from facts already observed, we should imagine that iron or zinc would calcine in water with greater facility than in vitriolic acid; and if some other circumstances did not interfere, it must be the case. This the following will in some degree illustrate.

Let S be a particle of sulphur, *d* a particle of dephlogisticated air, which it attracts with the force of 6 7-8ths, and let the compound be volatile sulphureous acid; here the tie between S and *d* is greater by 2-8ths, than that between the constituent principles of water, which is but 6 5-8ths.

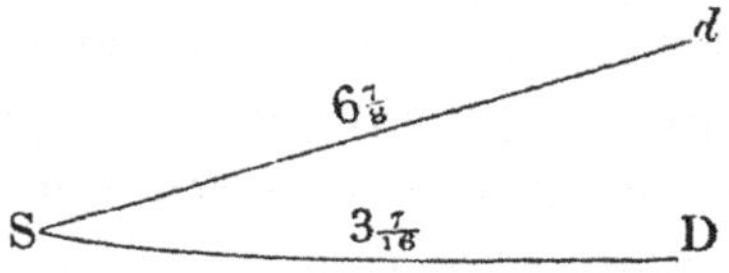

As the attraction of bodies is mutual, let us suppose S to possess one-half of this force, which is 3 7-16ths, and this to be its utmost exertion, and likewise *d* to possess the other half, which is 3 7-16ths more, which will unite them with the above-mentioned force. Let us suppose another particle of dephlogisticated air D to have a tendency to unite to S, with the force of 3 7-16ths, in order to form perfect vitriolic acid; to receive D, S must relax its attraction for *d* one-half. That is, the force of 3 7-16ths will be divided and directed in two different points, which will reduce the attachment of dephlogisticated air and sulphur in perfect vitriolic acid to 5 1-18th."

Page 66. "When vitriolic acid, whether diluted or not, is mixed with oil, an ultimate particle of vitriolic acid influences with a certain force an ultimate particle of oil, while the

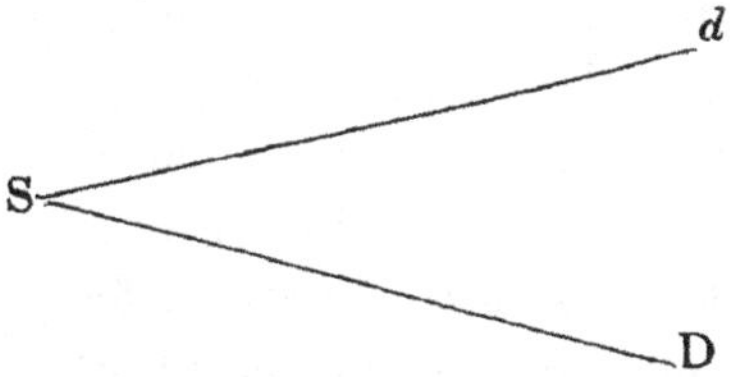

latter attracts the vitriolic acid with the same force. The oil will not take D *d* from S; but from the joint attraction of S —— D —— *d* to oil, they will approach with equal pace, and combine. Thus this mixture more than mechanically, but not quite chemically united, may be resolved into the different fluids mentioned above. The particle of oil will retain D or *d*, and form fixable air; at the same time that S will retain *d* or D with its full force, and form volatile vitriolic acid."

Page 132. "In my opinion the purest nitrous acid contains 5 of dephlogisticated to 1 of phlogisticated air. Nitrous air, according to Kirwan, contains 2 of dephlogisticated to 1 of phlogisticated air. According to Lavoisier, 100 grains of nitrous air contain 32 grains of phlogisticated air, and 68 of of dephlogisticated air. I am myself of the former philosopher's opinion; I likewise am of opinion, that every primary particle of phlogisticated air is united to two of dephlogisticated air, and that these molecules are surrounded with one common atmosphere of fire."

"To render this more explicable, let us suppose P to be an ultimate particle of phlogisticated air, which attracts dephlogisticated air with the force of 3; let *a* be a particle of dephlogisticated air, whose attraction to P we will suppose to be 3 more, by which they unite with the force of 6. The nature of this compound will be hereafter explained.

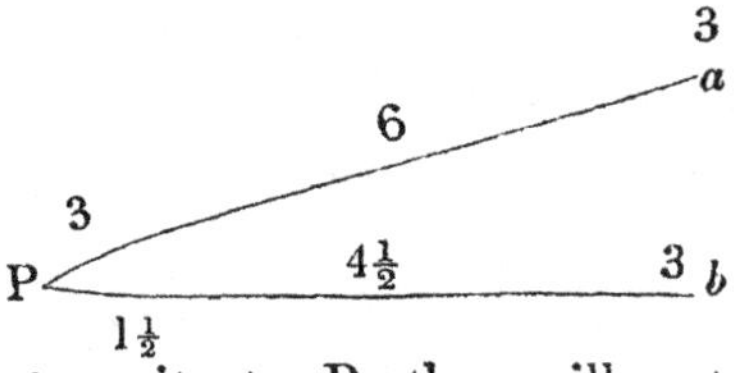

"Let us consider this to be the utmost force that can subsist between dephlogisticated and phlogisticated air. Let us suppose another particle of dephlogisticated air *b* to unite to P, they will not unite with the force of 6, but with the force of 4½; that is the whole power of P, which is but 3, will be equally divided and directed in two points towards *a* and *b*; so that P and *a b* will unite with the forces annexed to them; for the attraction of *a* and *b* to P meeting with no interruption, will suffer no diminution. This I consider to be the true state of nitrous air. Let us now suppose another particle of dephlogisticated air *c* to unite to P, it will combine only with the force of 4, whereby *a*, *b*, *c* and P will gravitate toward one another. Such is the state of the red nitrous vapour or the red nitrous acid.

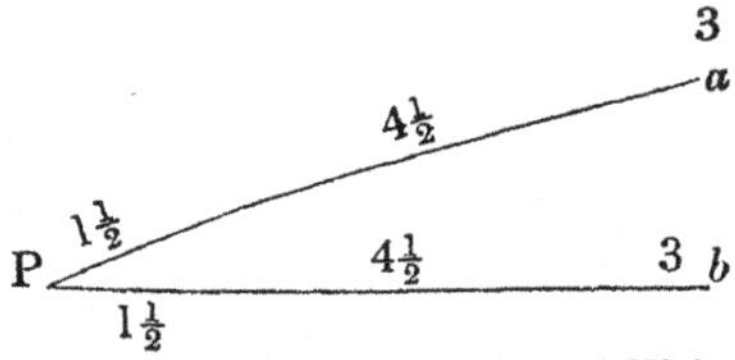

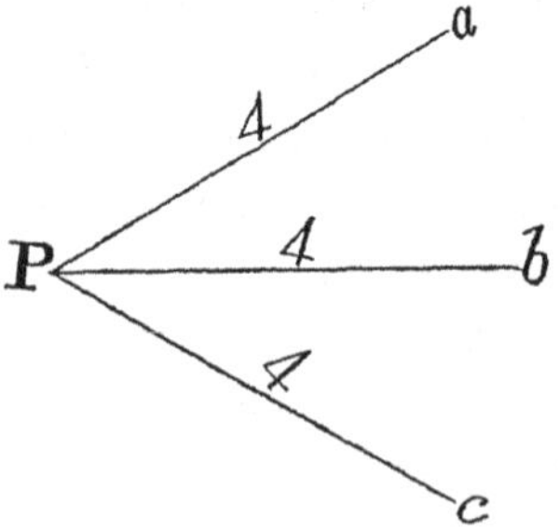

"Let us again suppose a fourth particle of dephlogisticated air *d* to combine with P, it will unite only with the force of $3\frac{3}{4}$. This, I think, is the state of the pale or straw coloured nitrous acid.

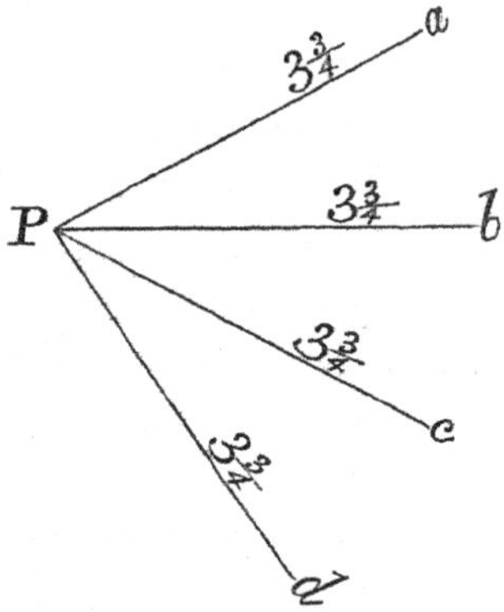

"Lastly, let us suppose a fifth particle of dephlogisticated air *e*, to unite to P, it will combine with the force of 3 3-5ths, so that *a*, *b*, *c*, *d* and *e* will each gravitate towards P as their common centre of gravity. This is the most perfect state of colourless nitrous acid; and in my opinion no more dephlogisticated air can unite to the phlogisticated air, as having its whole force of attraction expended on the particles of dephlogisticated air, *a*, *b*, *c*, *d*, *e*. *This illustrates the nature of saturation.* Thence we find that dephlogisticated air is retained with less force in the perfect or colourless nitrous acid, than in the straw-coloured, or in the red, or in nitrous air."

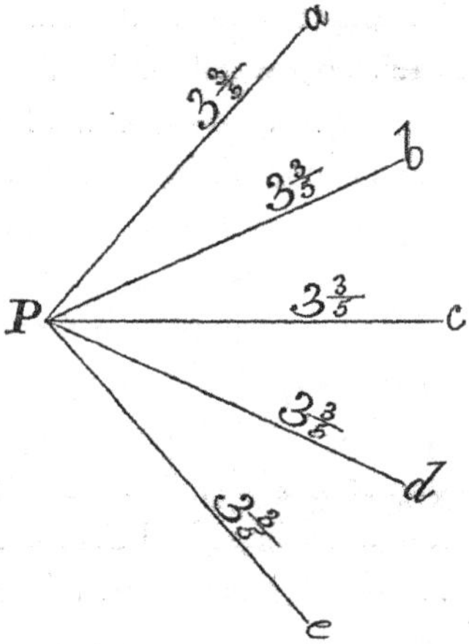

Page 194. "If the calcination of metals depended solely upon their union to dephlogisticated air, it must be supplied by water, when steam is brought in contact with them; and as every particle of light inflammable air is united but to a single particle of dephlogisticated air, inflammable air must be disengaged in proportion to the quantity of dephlogisticated air which unites to the metals; or, in other words, according to the degree of calcination it acquires."

Higgins wrote on phlogiston, his chief object being connected with it. In an octavo volume of above 300 pages these are nearly all the extracts relating to this subject. We find, then, that he did not establish a law in connection with these ideas. We must conclude, then, that he did not see their importance, that he did not see their application. Not only so, but being lost amongst so much material, we do not find that they were so written as to draw any attention, nor does he seem to have wished to do so. He wished to draw attention to what was in reality of less importance, his objections to phlogiston. The magic sword that would have slain all his enemies he threw away as some common truncheon. The magic lamp that would have given brilliance to him and to science, was found, after some years, rotting in his cellar. As regards the effect his book had on his cotemporaries, Dr. Thomson, of Glasgow, is the best authority. We see from him that no idea of an atomic theory was got from Higgins. His opinions were given and received as a speculation more than as expressing a fact or a law.

He says in his Annals of Philosophy, May, 1814, Vol. III., p. 331, "I have certainly affirmed that the atomic theory was not established in Mr. Higgins's book. And here is my reason. I have had that book in my possession since the year 1798, and had perused it carefully; yet I did not find anything in it which had suggested to me the atomic theory. That a small hint would have been sufficient I think pretty clear from this, that I was forcibly struck with Mr. Dalton's statements in 1804, though it did not fill half an octavo page; so much so, indeed, that I afterwards published an account of it; and I still consider myself as the first person who gave the world an outline of the Daltonian theory."

This is put too strongly. Had Dr. Thomson paid as much attention to Higgins's book as to the remarks of Dalton, he would certainly have made a great advance on the chemists of the period, by understanding definite proportion, and learning to reason in the spirit of the atomic theory.

William Higgins made an advance on Bryan Higgins in this theory of sulphur and heat, and he was a man evidently of an acute mind. But he was destined to find Emerson's saying true, that we often find in the sayings of great men our own rejected ideas. He was heir to the common opinion that atoms existed, and the opinion of Dr. Higgins that they united and formed molecules of compound bodies. He applied the reasoning further, and said that they must then unite in numbers of one or *two*, or three, and that there could be no intermediate combination, as there were no intermediate division of atoms. He applied this reason in two or three cases. These cases, such as nitric acid, are so clear and beautiful, that we can only be surprised that the general law was not seized on. They are the first clear and satisfactory reasons given for saturation, and for definite proportion in general. Higgins was therefore the first man who used the idea of atoms with such force as to be serviceable in chemistry. He used the idea of ultimate particles and the molecular state of bodies to illustrate saturation, and definite and multiple proportion, and gave us therefore the fundamental ideas of stoechiometry as they exists in chemical science, from which everything else might have easily flowed.

He had seen the right road, but dared not go farther. But we must take his own apology, "Est quoddam prodire tenus, si non datur ultra." It is something to have gone thus far, if he had no power to go beyond it.

Like Dr. Higgins, and the most of the chemists of last century, he assumed certain forces of attraction, and endeavoured to give comparative values to the forces of combination. These numbers representing affinity misled Higgins. Had he seen any general law, he would have seen that weights represented the comparative number of his atoms, but this he did not see with any distinctness, or he would have used Kirwan's numbers, which he knew well enough, and would have found in them atomic weights. But atomic weights were never thought of at all. It was not even certain then that all matters gravitated, although he held the present opinions on it. He reasoned to a certain extent in the true spirit of the atomic theory, but becoming entirely involved in dynamics, he entirely missed his way. Dynamics have hitherto entirely failed in chemistry. Their power has been used in upsetting their friends, and Higgins fell a victim to their forces.

Having for a moment laid hold of the idea that bodies which are atomically constituted must be formed of the union of one or more bodies, and of no intermediate atoms, he drew insufficient conclusions, and all its prospective advantages were lost to him. The want of suitable results, which it was his fault for not finding, seems to have caused him to let go his magic weapon, or to view his own opinion as a speculation. In his mind they exist as very little more.

I look upon him as the first man who ever in his imagination formed a correct atomic compound, and gave a correct analysis, in spite of the thousands of previous speculations and the simplicity of the idea, but one who lost the opportunity of elevating his idea into a great law of nature. It is well to express the claim of a discoverer in the widest and in the fewest words. He expressed the fact of atomic simple and multiple proportion, which is the foundation for all the other atomic laws, although in his mind it was not raised to the dignity of a great law, and it is for great laws only that we can give great honours in this case.

Higgins speaks so clearly and simply that we can readily believe that he would have illustrated the laws of chemical combination with great beauty had he seen the great value of his ideas. There is no obscurity in his language; there is no difficulty in telling exactly his place in science; but there is a difficulty in defining it exactly when we have to deal with Dalton, who grasped the whole so much more firmly, enlarged it, placed it, and established it in a series of laws. Any one to be put before Higgins must have made great advances: he cannot be put down by any obscure sentences dragged from any author. Any one to eclipse him must be fuller, more decisive, and more systematic.

From the want of these qualities Higgins appears more in the character of a great thinker than of a great discoverer.

CHAPTER IX.

RICHTER.

DURING the disputes as to Dalton's priority of discovery, it was frequently asserted that his atomic laws were not new, and they were, as is usual in such cases, attributed to various persons. Of these persons, Higgins, in this country, and Richter, in Germany, have been the most prominent. I have endeavoured to show exactly the position of Higgins; I shall do the same with Richter. Higgins came first with clearness and simplicity, uttering a beautiful idea which he failed to follow up; Richter came close after him, with great labour and enthusiasm, filled with a great idea of the study in which he was engaged, and obtained a law which he failed to follow up; he lost himself in complicated theories, having no idea how simple was the truth he sought for. Both were neglected, as happens when men fail to give completeness to their inquiries, even in the eyes of those who study and are willing to learn.

Richter's books are—"Anfangsgründe der Stoechyometrie oder Messkunst Chymischer Elemente." 3 vols. Bresslau und Hirschberg, 1792-4; and "Ueber die Neuern Gegenstände der Chymie," 1791-1802.

I shall give rather copious extracts from his works, shewing the direction of his inquiries, and the ends he attained.

RICHTER, VOL. I., PREFACE.

"Mathematics includes all those sciences which refer to magnitude, and consequently a science lies more or less in the province of mathematics (geometry), according as it requires the determination of magnitudes. In chemical experiments this truth has often led me to the question, whether and how far chemistry is a part of applied mathematics; and especially in considering the well-known fact, that two neutral salts, when they decompose each other, form again neutral compounds. The immediate consequence, in my opinion, could only be, *that there are definite relations between the magnitude* of the component parts of neutral salts. From that time I considered how these proportions could be made out, partly by exact chemical experiments, partly combining chemical with mathematical analysis. In my inaugural dissertation, published at Königsberg, in 1789, I made a slight attempt, but was not then supplied with the requisite chemical apparatus, nor was I sufficiently ready with all requisite information, bearing on my present system, imperfect as it may be. The result, therefore, was very imperfect. I promised, however, not to let the matter rest with that imperfect essay, but to work out this branch with all the accuracy and profundity of which I was capable, as soon as I was supplied with the requisite conveniences. This promise, I hope in the present volume, to make good, although I am far from believing that what I am now going to say will not be in need of still more thorough and accurate elaboration, for who will venture to limit the extent and the power which is the destination of a young and budding science."

He was the first to speak of a science of stœchiometry, and began formally to lay the foundation. We may even say that he commenced the systematic study for which he gave us also the most appropriate word. I cannot say that he began the science, and it will be seen that his mode of inquiry was wanting in directness and his results in completeness.

In page 29 of the preface, he says, "as the mathematical portion of chemistry deals in a great measure with bodies which are either elements or substances incapable of being decomposed and as it teaches also their relative magnitudes,

I have been able to find no more fitting name for this scientific discipline than the word stœchiometry, from στοιχειον which, in the Greek language, means a something which cannot be divided, and μετρειν which means to find out relative magnitudes."

Here, then, is a man prepared for the work, one who resolutely laboured for many years to find the law by which the elements combine, by "number, weight, and measure."

We have seen already that many facts were known, and that even reciprocal proportion was almost attained in the diagrams which have been given, and that the most far-sighted chemists saw the natural necessity for a constant proportion in combinations; but when the well-known laws agreed upon by chemists were put together, we see how few G. Morveau's list amounted to.

Richter did a great deal of work, especially in connection with the chemistry of the metals, but everything was held secondary to his great idea of definite proportionate quantities (bestimmte Grössenverhältnisse). On the title pages of the various papers or parts of volumes, written after his stœchiometry, he has preserved as a motto "Παντα (ΘΕΟΣ) μετρω και ἀριθμω και σταθμω διεταξε" (ας). This, from the 'Wisdom of Solomon,' chapter xi., v. 20, is exceedingly appropriate, but the context evidently shows that it was not applied to any such subject; at the same time it is introduced as a proverb would be, or a well-known universal law coming in aptly to illustrate one particular point to which it bears no more intimate relation than to innumerable others. It must, however, be confessed, that this expression is given with a minuteness and fulness which warrants the conclusion that it was not uttered until after many and profound speculations on the order of creation. The sentence is the expression of the circumstances in all their fulness, but like many other sentences of antiquity, the meaning is not clear till the facts have been discovered piece by piece.

The most important sentences bearing on the subject of Richter's volumes have been selected, including everything which seems to indicate any knowledge of the subject. His prolixity is excessive, every little idea is long dwelt upon, and as an example of the small fear he had of too much enlarging his book, it may be stated that he actually writes a system of algebra in one of the volumes, because a little algebra is wanted for the full understanding of his demonstrations. It may be that there are sentences hidden among other portions of the book less directly bearing on his subject which would indicate great knowledge, for although I have spent many days among his six volumes, I have certainly omitted some parts which seemed to me out of the range of stœchiometry. But his doctrines are not to be got in fragmentary sentences, so that the loss of any such sentences cannot, in the least, affect the result.

Richter's Stœchiometry, Vol. I., Page 121.

Definition 1.

"Stœchiometry (stœchyometria) is the science of measuring the quantitative proportions, or the proportions of the masses in which chemical elements stand in regard to each other. The mere knowledge of these relations might be called 'quantitative stœchiology.'

Principle 1. P. 123.

"Every infinitely small particle of the mass of an element has an infinitely small part of the chemical attractive force or affinity.

Experience 5.

"In order to make a neutral compound out of two elements, it is needful, as each of the elements is of the same constitution at one time as at another, to take the same quantity for the first part formed as for the second part. For example, if two parts of lime require five parts of muriatic acid for solution, six parts of lime will require fifteen of the same acid.

Experience 6. P. 124.

"When two neutral solutions are mixed, and a decomposition follows, the new resulting products are almost without exception neutral also, but if the solutions of one or both are not neutral before mixing, the products after mixture are also not neutral.

Corollary 1.

"The elements must therefore have amongst themselves a certain fixed proportion of mass. To determine which, their neutral compounds generally give the best opportunity.*

Corollary 2.

"If the weights of the masses of two neutral compounds which decompose each other are A and B, and the mass of the one element in A is *a*, and that of the one in B is *b*, then the masses of the elements in A are A—*a*, *a* and those in B are B—*b*, *b*. The proportions of the masses of the elements in the neutral compounds before decomposition are A—*a* : *a* and B—*b* : *b*; but after decomposition the new products are *a*+B—*b*, and *b*+A—*a*, and the proportion of the masses of the elements is *a* : B—*b*, *b* : A—*a*. If the proportion of the masses in the compounds A and B is known, that in the new products is known also.

* In German "Der Stoff ihrer neutraler Verbindungen öfters einen Bestimmungsgrund abgeben kann." Stoff is explained thus: Einleiting: Erklärung 14. "Das materielle oder körperliche Subjekt, worinnen sich die chymische Verwandtschaft befindet, nenne ich die *Masse, Prinzip* oder *Stoff (Massa)* des Elementes. Die Summa der Massen der Elemente, so eine neutrale auflösung bilden, ist die *Masse* oder *Stoff (Massa)* der neutralen Auflösung."

That is, "I call the material or corporal *subjectum* in which the chemical affinity resides, the Mass, Principle or Stuff (Massa)."

In a note, he says,—"There is present in the Element a certain *subjectum* to which the chemical attractive power or the affinity is bound, this is the Mass of the Element."

If $a+B-b=C$ and $b+A-a=D$ then $a=C+b-B=b+A-D$ and $C-B=A-D$, so also $D-B=A-C$. In addition $b=a+B-C=D-A+a$.

THEOREM 1. P. 125.

"The chemically attracting power by which one element *a* enters into neutrality with another A—*a* presupposes an opposite action of the same kind in the latter, and these two powers are equal to each other.

THEOREM 2. P. 128.

"If a neutral compound A whose elementary masses A—*a* and *a* are removed from combination by a definite quantity of a third element *b*, and the whole mass of one element *a* for example is set free, the force that causes this phenomenon is equal to the difference between the separating element *b* and the separated element *a*.

THEOREM 3. P. 130.

"When two neutral compounds A and B, the masses of whose ingredients are A—*a*, *a* and B—*b*, *b* mutually decompose each other, so that the new products A—*a*+*b* and B—*b*+*a* are formed, the forces that partly cause and partly hinder this action, are equal to the difference of affinities of the elements A—*a* and B—*b* towards each of the elements *a* and *b*." Afterwards he said,

"When I finished the pure stœchiometry, two years ago, I did not think it would be needful to make any additions to its contents. In the first place I thought it had all that practical stœchiometry required, &c.*

VOL. II., PAR. V., P. 4.

Having decided by experiment that 1000 parts carbonate of lime contain 559 earthy matter, he sums up as follows

* Preface to Vol. I., Part 2. 1794. Later than Vols. II. and III.

his experiment with 5760 grs. of muriatic acid and 2393 grs. of chalk.

"If now we desire to find the proportion of the elements in the pure salt forming a neutral body, we must first seek to determine the amount of lime out of the weight of the crude lime used or the aerial salt of lime. This amounts to 2393 grains. According, then, to par. 1, 1000:559=2393 : lime, and the lime is equal to $\frac{2393 \cdot 559}{1000}$=1337; this, when subtracted from the 2544 grains of the neutral mass obtained, leaves a residue of 1207 grains, the weight of the muriatic acid. If, then, 1207 : 2544*=1000 : 1107, it is clear that in the salt of lime (chloride of calcium) 1000 parts of muriatic acid are united in a neutral state with 1107 parts of lime; the proportion of the elements in this neutral solution is then best designated by 1000 : 1107.

In this manner all the earths are treated, after which he gives the relation of the quantities of alkaline earths towards sulphuric acid and each other.

Order of the masses of alkaline earths towards muriatic acid. § XXII. P. 27.

"If we set in a row the numbers which have been found representing the masses of alkaline earths which unite with 1000 parts of muriatic acid, we obtain the first series of quantities of the alkaline earths. The muriatic acid is the determining element (elementum determinans) of this series, and every member of this series represents an element determined (elementum determinatum). In order to designate the elements to which we affix these numbers, we shall make use of the chemical signs for the sake of convenience, setting the determining element, or rather its sign, at the top, or at the side of the series of quantities; and when no number is placed, we shall suppose it to be 1000. In order to fix these signs in our memory, we shall here repeat them." (These signs it is not convenient to use.)

* This evidently ought to have been 1337.

"According to the paragraphs quoted, the following is the series of the alkaline earths in their relation to muriatic acid:—

MURIATIC ACID.

ALUMINA.	MAGNESIA.	LIME.	BARYTA.
734	858	1107	3099

"Little as the members of this series appear to follow any certain order, it is nevertheless decidedly the case; at the same time the inquiry into the law of these series is one of the most difficult problems which stœchiometry gives us to solve, and if we do not go to the inquiry with sufficient practical and theoretical exactness, we shall not succeed in our inquiry into these laws or orders (of the numbers). And now, to inquire into the law of the series before us, let us first seek the difference between each member and its successor, and we obtain 858—734=124, 1107—858=249, 3099—1107=1992. Let us then use the first difference to divide the two following differences, and we obtain $\frac{249}{124}=2+\frac{1}{124}$, $\frac{1992}{124}=16\frac{8}{124}$. Then let us see if one quotient allows of division by the other, that is, let us divide $16+\frac{8}{124}$ by $2+\frac{1}{124}$. If we bring divisor and dividend under the same denomination of 124, this will be $16+\frac{8}{124}=\frac{1992}{124}$ and $2+\frac{1}{124}=\frac{249}{124}$; therefore $\frac{1992:124}{249:124}=\frac{1992}{249}$ and 249 is contained exactly 8 times in 1992, consequently $\frac{1992}{249}=8$. From this it is clear, that when we have the first difference $124\frac{1}{2}=\frac{249}{2}$, all the succeeding differences may be so divided by it that nothing remains; the half here mentioned is only $\frac{1}{1716}$ in 858 parts=0.0006 and still less in the other members of the series, it is, therefore, of no importance; it is impossible in experiments to arrive at such minuteness, at the same time in calculating the proportion to 1000 parts, it was necessary to throw away small unimportant fractions, otherwise it would be needful to use an enormous number of figures in order to designate the quantities. Now $\frac{249}{2}$ X 2=249; and $\frac{249}{2}$ X 8=1992, consequently $734+\frac{249}{2}=858\frac{1}{2}$, $734+\frac{249}{2}+\frac{249}{2}$ X 2=$1107\frac{1}{2}$; 734

2 C

$+\frac{249}{2}+\frac{249\times 2}{2}+\frac{249}{2}\times 16=3099\frac{1}{2}$. In order better to understand all, let us make $734=a$, $\frac{249}{2}=b$, then $734=a$, $858\frac{1}{2}=a+b$, $1107\frac{1}{2}=a+b+2b=a+3b$, $3099\frac{1}{2}=a+b+2b+16b=a+19b$. From this the quantitative series appears in the following order:—

MURIATIC ACID.

ALUMINA.	MAGNESIA.	LIME.	BARYTA.
a	$a+b$	$a+3\,b$	$a+19\,b$

"Now this series remains always the same, even when we put a higher or a lower number for the mass of the determining element, for if the mass of the determining element is *n* times greater or *n* times smaller, then, in the first case, all the terms would be *n* times greater; that is, multiplied by *n*; and in the latter case *n* times smaller or divided by *n*, and the order of the differences would remain always the same; because what occurs with one of the differences must occur also with the others, if otherwise the determining element must still be considered as such. When this series is attentively considered, we observe that the difference of the successive terms is a mathematical product of the first difference *b* with an odd number. According to it the quantities in which the hitherto known alkaline earths assert their neutrality with muriatic acid are terms of a real arithmetical progression, the terms of which are found, when the product of a certain quantity with an odd number is added to the first term, only that between them many odd numbers, such as 5, 7, 9, 11, 13, 17, are left out. This is more remarkable, as the differences which the first term makes with the succeeding ones may be represented entirely by odd numbers; for one need only suppose that the mass of the determining element is divided by *b*, then all the terms of the series would be at once divided by *b*, and appear in the following form:—

$$\frac{\text{Muriatic Acid.}}{b}=\frac{1000}{249:2}=\frac{2000}{249}=8\frac{8}{249}$$

ALUMINA.	MAGNESIA.	LIME.	* * *	BARYTA. . . . .
$\frac{a}{b}$	$\frac{a}{b}+1$	$\frac{a}{b}+3$	* * *	$\frac{a}{b}+19$ . . . .

"In this case the first term would be $\frac{a}{b}$ and if it were all expressed in numbers then would $\frac{a}{b}=\frac{734}{249:2}=\frac{1468}{249}=5+\frac{223}{249}$ and the mass of the determining element $\frac{1000}{b}=\frac{1000}{249:2}=\frac{2000}{249}=8\frac{8}{249}$. In this way all the members are obtained in numbers, when 1, 3, and 19 are added to the first term $\frac{1468}{249}$, and the elements observed which are designated by these figures. It is very probable that the terms $\frac{a}{b}+5$, $\frac{a}{b}+7$, $\frac{a}{b}+9$, $\frac{a}{b}+11$, $\frac{a}{b}+13$, $\frac{a}{b}+15$, $\frac{a}{b}+17$ are wanting in the series, and the reasons for considering this probable, will be shewn in a suitable place.

"*Preliminary determination of the order of the alkaline earths which enter into neutrality with vitriolic acid:* § XXIII., p. 33.

"If we put in order the amount (mass) of alkaline earths which stand in neutrality with 1000 parts of vitriolic acid, in the manner adopted with muriatic acid, the following series of quantities is obtained:—

VITRIOLIC ACID.

MAGNESIA.	LIME.	ALUMINA.	BARYTA.
616	796	1053	2226

"In order to discover the law of this series let us, as in the former case, subtract the first term from all the succeeding, and we receive 796—616=180; 1053—616=437, 2226—616 =1610. Let us see now if the first difference can so divide all the rest, that nothing, or at least very little, remains, then $\frac{437}{180}=2+\frac{77}{180}$, $\frac{1610}{180}=8+\frac{170}{180}$. As nothing can be discovered here on account of the variety in the remaining fractions, let us divide every difference by 90 as the half of the first difference, and we obtain $\frac{180}{90}=2$, $\frac{437}{90}=5$, $\frac{1610}{90}=18-\frac{10}{90}$. The fractions here are not so considerable as before, although too large to be thrown away. Until, therefore, we are able to complete the order let us make 616=616, 796=616—2.90,* 1053=616+5.90—$\frac{13}{90}$, 2226=616+18.90—$\frac{10}{90}$."

* 2.90 means 2 × 90.

"*Nearer determination of the law by which the quantities of the alkaline earths, which enter into rest and neutrality with muriatic and vitriolic acid, increase or diminish in arithmetical progression.* § XXIV., p. 34.

"A. As we cannot completely ascertain the law by which the terms of the two series of numbers obtained by experiment proceed, we must try another source of information, to the obtaining of which the series itself which the determining element of muriatic acid makes with the alkaline earths, gives us an opportunity. As the differences of the quantities in § XXII. are a product of a quantity b with an odd number, it is possible that as many terms are wanting as there are odd numbers between 3 and 19, and even that other terms may lie beyond the term $a+19b$ or $\frac{a}{b}+19$. Suppose, then, that this series were complete, namely, a, $a+b$, $a+3b$, $a+5b$, $a+7b$, $a+9b$, $a+11b$, $a+13b$, $a+15b$, $a+17b$, $a+19b$, $a+21b$, $a+23b$, &c., the masses of the elements which enter into neutrality with 1000 parts of muriatic acid would be the following:—

a			=	734			=	734
a	+	b	=	734	+	$124\frac{1}{2}$	=	$858\frac{1}{2}$
a	+	$3b$	=	734	+	$3.124\frac{1}{2}$	=	$1107\frac{1}{2}$
a	+	$5b$	=	734	+	$5.124\frac{1}{2}$	=	$1356\frac{1}{2}$
a	+	$7b$	=	734	+	$7.124\frac{1}{2}$	=	$1605\frac{1}{2}$
a	+	$9b$	=	734	+	$9.124\frac{1}{2}$	=	$1854\frac{1}{2}$
a	+	$11b$	=	734	+	$11.124\frac{1}{2}$	=	$2103\frac{1}{2}$
a	+	$13b$	=	734	+	$13.124\frac{1}{2}$	=	$2352\frac{1}{2}$
a	+	$15b$	=	734	+	$15.124\frac{1}{2}$	=	$2601\frac{1}{2}$
a	+	$17b$	=	734	+	$17.124\frac{1}{2}$	=	$2850\frac{1}{2}$
a	+	$19b$	=	734	+	$19.124\frac{1}{2}$	=	$3099\frac{1}{2}$
a	+	$21b$	=	734	+	$21.124\frac{1}{2}$	=	$3348\frac{1}{2}$
a	+	$23b$	=	734	+	$23.124\frac{1}{2}$	=	$3597\frac{1}{2}$
		&c.			&c.		&c.	

"Now let us suppose that these alkaline earths, which are partly real, partly possible, and designated by the above numbers, entered with muriatic acid into such neutral combinations as would decompose with a neutral compound out of the series § XXIII., such as the magnesia salt, by double affinity either positive or negative (see Pure Stœchiometry. Theor. 3, coroll. 3. Introd. definition 16), then only that neutral compound is excepted which the muriatic acid makes with the alkaline element of the neutral salt which we have chosen in the series XXIII., or the combination which is also taken with the magnesia salt. But according to experiment 6, coroll. 2, in the Pure Stœchiometry, in the decomposition by double affinity, out of three proportions the fourth may be determined. Let us suppose, then, that all the mentioned actual and possible neutral combinations, magnesia salt excepted, decompose with sulphate of magnesia (bitter salt) either positively or negatively, so that each constituent is placed in a state of rest (Pure Stœch. Theor. 1, coroll. 1); then we may find how many measures of each of the real and possible elements are wanted for 1000 parts of the vitriolic acid measures. (Pure Stoech. Introd. def. 14.)

"B. The first neutral compound in the series, § XXII., is an actual one, namely, alum salt, where 734 parts of alumina stand in neutrality with 1000 parts of muriatic acid. If this neutral or middle salt decomposes with sulphate of magnesia by double affinity, then 858½ parts of magnesia must be contained in the sulphate of magnesia, because the proportionate quantity in the magnesia salt is as 1000 : 858, or rather 1000 : 858½. § XXII. Now the proportionate quantity in the sulphate of magnesia is 1000 : 616 (§ XIX., XXIII.) and 616 : 1000=858½ : 1,394; that is, if 616 parts of magnesia insist on rest with 1000 parts of vitriolic acid, the same must occur between 858½ parts of the first and 1394 parts of the latter; therefore when alumina salt and magnesia salt decompose, 1000 parts of muriatic acid dissolve with 858½ parts of magnesia, and 1394 parts of vitriolic acid with 734 parts of alumina; the proportionate amount of the alum formed will then be 1394 : 734=1000 : 526, which is not the proportion of the neutral but the common alum. (§ XXI.) The quantity of the alkaline earth in common alum belongs accordingly to the series § XXIII.

"C. But when, in the decomposition of the first neutral compound, where muriatic acid is the determining element with sulphate of magnesia, the amount of the vitriolic acid is 1394, it is so in all subsequent possible decompositions which the neutral compounds of the other actual and possible elements of the series § XXII. make with the sulphate of magnesia, let these decompositions be positive or negative. (Pure Stœch. Theor. 3, coroll. 3.) The quantity, then, of real and possible elements which belong to 1000 parts of muriatic acid belong to 1394 of vitriolic, and the following proportions are obtained for the compounds where the vitriolic acid becomes at rest with the real and possible elements (Pure Stoech. Theor. 1, coroll. 1), all of which obtained by experiment except the common alum are neutral.

1394	:	734	=	1000	:	526
1394	:	858½	=	1000	:	616
1394	:	1107½	=	1000	:	796
1394	:	1356½	=	1000	:	973
1394	:	1605½	=	1000	:	1152
1394	:	1854½	=	1000	:	1330
1394	:	2103½	=	1000	:	1508
1394	:	2352½	=	1000	:	1687
1394	:	2601½	=	1000	:	1866
1394	:	2850½	=	1000	:	2045
1394	:	3099½	=	1000	:	2224
1394	:	3348½	=	1000	:	2402
1394	:	3597½	=	1000	:	2580

§ XXV.

" A. When we look on all these numbers found, viz., 526, 616, 796, 973, &c., as quantities of the elements which are at rest with 1000 parts of sulphuric acid, we obtain a series, the law of which soon appears to us. Let us first subtract the first term from all the succeeding, and we obtain the following differences, which may be expressed in various ways :—

616	—	526	=	90	=	90			=	90		
796	—	526	=	270	=	270			=	3.90		
973	—	526	=	447	=	450	—	3	=	5.90	—	3
1152	—	526	=	626	=	630	—	4	=	7.90	—	4
1330	—	526	=	804	=	810	—	6	=	9.90	—	6
1508	—	526	=	982	=	990	—	8	=	11.90	—	8
1687	—	526	=	1161	=	1170	—	9	=	13.90	—	9
1866	—	526	=	1340	=	1350	—	10	=	15.90	—	10
2045	—	526	=	1519	=	1530	—	11	=	17.90	—	11
2224	—	526	=	1698	=	1710	—	12	=	19.90	—	12
2402	—	526	=	1876	=	1890	—	14	=	21.90	—	14
2580	—	526	=	2054	=	2070	—	16	=	23.90	—	16

" B. The law by which the differences of the actual and possible alkaline earths increase in relation to vitriolic acid is then so far made out that it follows from the product of a number which is here 90 with the consecutive odd numbers; we may consider the numbers which are to be subtracted, such as 3, 4, 6, 8, 9, &c., as nothing, because the greatest error that could be caused is only $\frac{3}{450}=\frac{1}{150}=0.0066$, or $\frac{66}{10000}$; but even this is not necessary, because the numbers themselves proceed in distinct order, as the series shews; and if we inquire, as in § XXII., into the manner in which these numbers progress, we observe that if three of them increase by odd numbers, the succeeding four increase in the ordinary way by one, and so alternately; for example.

—	3			=	—	3
—	(3	+	1)	=	—	4
—	(3	+	3)	=	—	6
—	(3	+	5)	=	—	8
—	(3	+	6)	=	—	9
—	(3	+	7)	=	—	10
—	(3	+	8)	=	—	11
—	(3	+	9)	=	—	12
—	(3	+	11)	=	—	14
—	(3	+	13)	=	—	16
—	(3	+	15)	=	—	18
—	(3	+	16)	=	—	19
		&c.		&c.		

"These quantities accordingly proceed in arithmetical progression down to the most insignificant fractions, as we may see by a glance at them.

"*The quantities in which the alkaline earths enter into neutrality with muriatic acid are terms of an endless series, which increase by the product of a determinate quantity with the consecutive odd numbers. The same thing occurs with the alkaline earths in relation to sulphuric acid, only that in this case a quantity must be taken from the terms of the last series, the first three excepted; this quantity increasing also in progression.*

§ XXVI.

"A. After finding out the law by which the quantities of the alkaline earths increase towards the two acids (sulphuric and muriatic), it becomes necessary to form the series themselves, that we may see clearly the correctness of the proposition advanced as a hypothesis; for if it is done rightly the proposition ceases to be a hypothesis. We shall designate the terms that are wanting in both series by a star, and the elements which produce with acids a very violent heat when freed from their air, for example, lime and magnesia, by △, as the sign of fire.

Muriatic Acid .... $a = 734$, $b = \frac{249}{2} = 124\frac{1}{2}$.

Alumina...	a			=	734	+		=	734
△ Magnesia..	a	+	b	=	734	+	$\frac{249}{2}$	=	$858\frac{1}{2}$
△ Lime	a	+	$3b$	=	734	+	$\frac{3 \cdot 249}{2}$	=	$1107\frac{1}{2}$
*	a	+	$5b$	=	734	+	$\frac{5 \cdot 249}{2}$	=	$1356\frac{1}{2}$
*	a	+	$7b$	=	734	+	$\frac{7 \cdot 249}{2}$	=	$1605\frac{1}{2}$
*	a	+	$9b$	=	734	+	$\frac{9 \cdot 249}{2}$	=	$1854\frac{1}{2}$
*	a	+	$11b$	=	734	+	$\frac{11 \cdot 249}{2}$	=	$2103\frac{1}{2}$
*	a	+	$13b$	=	734	+	$\frac{13 \cdot 249}{2}$	=	$2352\frac{1}{2}$
*	a	+	$15b$	=	734	+	$\frac{15 \cdot 249}{2}$	=	$2601\frac{1}{2}$
*	a	+	$17b$	=	734	+	$\frac{17 \cdot 249}{2}$	=	$2850\frac{1}{2}$
Baryta ..	a	+	$19b$	=	734	+	$\frac{19 \cdot 249}{2}$	=	$3099\frac{1}{2}$
*	a	+	$21b$	=	734	+	$\frac{21 \cdot 249}{2}$	=	$3348\frac{1}{2}$
*	a	+	$23b$	=	734	+	$\frac{23 \cdot 249}{2}$	=	$3597\frac{1}{2}$
			&c.				&c.		

" B. Before we set down the quantitative progression in the case of vitriolic acid, we must first inquire if the quantity of alumina in neutral alum belongs to this series; it is 1053. Let us subtract 526 from 1053, and we obtain 527; now 527= 540—13=6.90—13, and consequently 1053=526+6.90—13. But as the series determined by vitriolic acid proceeds by the uninterrupted odd numbers, and no neutral alum can be found in decomposing by double affinity, the quantity 526+6.90—13 does not belong to this series. We must take it, however, in the meantime into the series, because it belongs to the quantities which enter into neutrality. We shall, however, put such in brackets, as must happen when considering the quantity of alumina in common alum, if it is not a legitimate member of the series, and capable of double affinity.

No. 2.

Sulphuric Acid......$a = 526$, $b = 90$

	Alumina..	a	$=526$	$=526$
△	Magnesia.	$a+b$	$=526+90$	$=616$
△	Lime . . .	$a+3b$	$=526+3.90$	$=796$
	*	$a+5b-3$	$=526+5.90-3$	$=973$
	Alumina..	$(a+6b-13)$	$=526+(6.90-13)$	$=1053$
	*	$a+7b-(3+1)$	$=526+7.90-(3+1)$	$=1152$
	*	$a+9b-(3+3)$	$=526+9.90-(3+3)$	$=1330$
	*	$a+11b-(3+5)$	$=526+11.90-(3+5)$	$=1508$
	*	$a+13b-(3+6)$	$=526+13.90-(3+6)$	$=1687$
	*	$a+15b-(3+7)$	$=526+15.90-(3+7)$	$=1866$
	*	$a+17b-(3+8)$	$=526+17.90-(3+8)$	$=2045$
	Baryta . .	$a+19b-(3+9)$	$=526+19.90-(3+9)$	$=2224$
	*	$a+21b-(3+11)$	$=526+21.90-(3+11)$	$=2402$
	*	$a+23b-(3+13)$	$=526+23.90-(3+13)$	$=2580$
		&c.	&c.	

" C. If we convert the differences into simple consecutive odd numbers, the quantity of the determining element and all the terms of the series have only to be divided by b, and we receive:

No. 1.

$\frac{\text{Muriatic Acid},}{b}$ $a=734$, $b=\frac{249}{2}$ $\frac{\text{Muriatic Acid},}{b}=8\frac{+8}{249}$

	Alumina..	$\frac{a}{b}$	$=\frac{734}{124\frac{1}{2}}$	$=5+\frac{223}{249}$
△	Magnesia.	$\frac{a}{b}+1$	$=\frac{858\frac{1}{2}}{124\frac{1}{2}}$	$=6+\frac{223}{249}$
△	Lime....	$\frac{a}{b}+3$	$=\frac{1107\frac{1}{2}}{124\frac{1}{2}}$	$=8+\frac{223}{249}$
	*	$\frac{a}{b}+5$	$=\frac{1356\frac{1}{2}}{124\frac{1}{2}}$	$=10+\frac{223}{249}$
	*	$\frac{a}{b}+7$	$=\frac{1605\frac{1}{2}}{124\frac{1}{2}}$	$=12+\frac{223}{249}$
	*	$\frac{a}{b}+9$	$=\frac{1854\frac{1}{2}}{124\frac{1}{2}}$	$=14+\frac{223}{249}$
	*	$\frac{a}{b}+11$	$=\frac{2103\frac{1}{2}}{124\frac{1}{2}}$	$=16+\frac{223}{249}$

$* \quad \frac{a}{b} + 13 = \frac{2352\frac{1}{2}}{124\frac{1}{2}} = 18 + \frac{223}{249}$

$* \quad \frac{a}{b} + 15 = \frac{2601\frac{1}{2}}{124\frac{1}{2}} = 20 + \frac{223}{249}$

$* \quad \frac{a}{b} + 17 = \frac{2850\frac{1}{2}}{124\frac{1}{2}} = 22 + \frac{223}{249}$

Baryta.. $\frac{a}{b} + 19 = \frac{3099\frac{1}{2}}{124\frac{1}{2}} = 24 + \frac{223}{249}$

$* \quad \frac{a}{b} + 21 = \frac{3348\frac{1}{2}}{124\frac{1}{2}} = 26 + \frac{223}{249}$

$* \quad \frac{a}{b} + 23 = \frac{3597\frac{1}{2}}{124\frac{1}{2}} = 28 + \frac{223}{249}$

&c. &c.

No. 2.

Further—

Vitriolic Acid, $a=526$, $b=90$ $\quad \frac{\text{Vitriolic Acid}}{b} = 11\frac{1}{3}$

Alumina. $\frac{a}{b} = \frac{526}{90} = 5 + \frac{76}{90}$

Magnesia $\frac{a}{b} + 1 = \frac{616}{90} = 6 + \frac{76}{90}$

Lime .. $\frac{a}{b} + 3 = \frac{796}{90} = 8 + \frac{76}{90}$

$* \quad \frac{a}{b} + 5 - \frac{3}{b} = \frac{973}{90} = 10 + \frac{73}{90}$

Alumina. $\left[\frac{a}{b} + 6 - \frac{13}{b}\right] = \left[\frac{1053}{90}\right] = \left[11 + \frac{63}{90}\right]$

$* \quad \frac{a}{b} + 7 - \frac{(3+1)}{b} = \frac{1152}{90} = 12 + \frac{72}{90}$

$* \quad \frac{a}{b} + 9 - \frac{(3+3)}{b} = \frac{1330}{90} = 14 + \frac{70}{90}$

$* \quad \frac{a}{b} + 11 - \frac{(3+5)}{b} = \frac{1508}{90} = 16 + \frac{68}{90}$

$* \quad \frac{a}{b} + 13 - \frac{(3+6)}{b} = \frac{1687}{90} = 18 + \frac{67}{90}$

$* \quad \frac{a}{b} + 15 - \frac{(3+7)}{b} = \frac{1866}{90} = 20 + \frac{66}{90}$

$* \quad \frac{a}{b} + 17 - \frac{(3+8)}{b} = \frac{2045}{90} = 22 + \frac{65}{90}$

Baryta .. $\frac{a}{b} + 19 - \frac{(3+9)}{b} = \frac{2224}{90} = 24 + \frac{64}{90}$

$* \quad \frac{a}{b} + 21 - \frac{(3+11)}{b} = \frac{2402}{90} = 26 + \frac{62}{90}$

$* \quad \frac{a}{b} + 23 - \frac{(3+13)}{b} = \frac{2580}{90} = 28 + \frac{60}{90}$

&c. &c.

"D. When the numbers in the last series are compared with those found by experiment, they are found to agree perfectly, as far as regards alumina, lime, and magnesia. On the other hand, the amount of baryta in the sulphate is 2224, but in § XIX. 2226. No doubt the difference comes from the greater difficulty in finding the point of saturation in the case of this earth than in the case of the others, when combining them with sulphuric acid. At the same time, the supposed error is so small that it may be left out of consideration, for it amounts only to $\frac{2}{2224}$ or 0.0009 or $\frac{9}{10000}$, which is a difference that may be reckoned as nothing. It must be remembered that decimal fractions only are used here.

"E. Now if the quantities found by experiment exactly fit into the series, if all the terms of these series entirely correspond to the possibility of double affinity, if even a quantity which is capable of neutrality, but not of double affinity, is banished out of one series by the rule of the series; further, if one series becomes possible only through the other, the proposition is absolutely certain, that the quantities of the hitherto known alkaline earths which enter into rest or equilibrium with sulphuric and muriatic acids are terms of an infinite series in arithmetical progression, each of which proceeds according to its own law.

"G. Shall we then conclude from the laws of the two series, in which so many terms are wanting, that there are many alkaline earths existing in nature? So far as probability and possibility are concerned it is a fair conclusion, especially since the knowledge of magnesia and baryta are the property of only the last half century. If they had not been discovered until after the study of the stœchiometric sphere had commenced, the second and eleventh term of every series would be wanting, and a * substituted, and besides it would not have been possible, with such a small number of terms to have found out the law. Who knows if there are not other elements in existence which interrupt the last series as neutral quantities, precisely as with the alumina? But if we should conclude from the law of these series that the existence of the failing elements is necessary, then we should commit as great an error as if we were to conclude that a planet must exist between Mars and Jupiter, because it corresponds to the law of the distance of the planets from the sun.

"H. The use of these series of quantities is not small, for if we know only the first member and are acquainted with the law, we find all the other members and all the proportionate quantities with the greatest exactness, and who does not know what differences there have hitherto been in the numbers representing the proportions? How many uses shall we find also in chemical analysis for series of this kind, of which probably there are many, and to what perfection might it not bring the chemical system, if they could be used as tables of affinities?

"*The two series of quantities,* 1 *and* 2, § XXVI., *are really quantitative series of the affinity of alkaline earths towards muriatic and sulphuric acid.* Page 51.

* * * * * * * * *

Page 56. "*Determination of the decomposing forces,* § XXVIII.

"A. The experiments now detailed enable us to state the proposition that affinities are as the masses.

"C. P. 61. If now in these cases of affinity the attracting forces of the elements are as the masses of the elements, and we take 3099 as the attracting force or affinity of baryta towards muriatic acid, the numbers 1107, 858, 734, or the attractive force of the elements represented is quite unaltered towards muriatic acid, on the other hand we must calculate the affinity of these earths to sulphuric acid from the numbers given and the proposition adopted. According to the proposition, 1000 : 1394=3099 : 4320 and 1000 : 1394=734 : 1023, in the same way 3099 : 734= 4320 : 1023. If, then, the attractive power of baryta towards muriatic acid is 3099, towards vitriolic acid it is 4320, and if alumina acts towards muriatic acid with the force of 734, and towards vitriolic acid with a force of 1023 forming common alum, in the same manner the baryta is attracted with a force of 4320 towards vitriolic acid, forming heavy spar, and so the power by which this acid forms common alum is only 1023. If we inquire into the affinity of the other alkaline earths towards vitriolic acid by the rule of three, we obtain for lime $\frac{1394 \cdot 1107}{1000}=1543$, for magnesia $\frac{1394 \cdot 858}{1000}=1196$. If, then, in the cases of double affinity given, we put instead of the quantity the attractive force by which one element works on another, we obtain, according to the first theorem of the 'Pure Stœchiometry,' as follows:—

No. 1.

Baryta.		Muriatic Acid.
3099	Baryta Salt.	3099
4320		734
	Heavy Salt.	
	Alum. Spar.	
734		4320
1023	Common Alum.	1023
Alumina.		Vitriolic Acid.

"D. * * * In No. 1 the two positive or decomposing elements are 4320 and 734, the negative which hinder the compound are 3099 and 1023, consequently 4320+734=5054 the whole positive or furthering, and 3099+1023 =4122 the whole negative or hindering power. * * *" The difference (equal to the power, as he says) =+932 is positive.

He then endeavours to shew in the same way, page 171, &c., that "The masses (quantities) of the three alkaline salts, which enter into neutrality with an equal amount of vitriolic or muriatic acid, are the three first terms of two series, of which, that which belongs to muriatic acid proceeds by the odd numbers without interruption, and the other is the product of a quantity with the numbers in regular succession."

Page 167. "When an aqueous solution of vitriolic salammoniac is poured into a solution of muriate of lime, an abundant precipitate is caused, which is completely formed gypsum; if the exact quantity of the salammoniac solution has been used which is necessary to complete the precipitation, the liquid above the precipitate contains nothing but perfectly formed common salammoniac. But the proportion in the salt is 1000 : 1107½, and to 1000 of the chloride are to be calculated 889 parts of the volatile alkali; now let us inquire how much of the vitriol is needed for 1107½ parts of the lime, the proportion of the last to the first is 796 : 1000, consequently 1107½ parts of lime demand $\frac{1000 \cdot 1107\frac{1}{2}}{796} = \frac{2215000}{1592} = 1394$ parts of vitriol, which belong to the 889 parts of the volatile alkali.

He gives a list, page 279, of "*Proportional quantities of neutral compounds which decompose each other, when entirely deprived of water.*" * * * * "In each of these cases it is only necessary to add the numbers representing the quantities standing against each other horizontally, by which the power of the affinity is estimated, and we receive the neutral quantities which decompose each other, and consequently their proportion."

Salammoniac Vitriol.		Common Salt.
689 + 1000	:	960 + 717 = 1689 : 1677
Salammoniac Vitriol.		Common Salt.
638 + 1000	:	960 + 717 = 1638 : 1677
Magnesia Salt.		Vitriolized Potash.
858 + 1000	:	2239 + 1394 = 1858 : 3633

&c., &c., &c.; this is from a list of 28: the second and third are supposed hydrous.

Then we have, p. 284, "*Proportional quantities of neutral compounds containing muriatic acid, considered as anhydrous, when decomposed by vitriolic acid.*" Also, "*Proportional quantities, when the neutral compounds which vitriolic acid makes with the alkaline salts and magnesia are decomposed negatively or by free muriatic acid.*" P. 293.

At page 190, he says, that the affinities are as the amount of the combining proportions, and here also the atomic weights of ammonia, soda, and potash, are such as to lend some countenance to it. The series, however, is still considered the most important thing, and he finds afterwards, in the vol. for 1800, that smaller weights may precipitate larger ones.

These inquiries were continued with great labour, and in his work "On the newer subjects in chemistry," we have many attempts to define the relations between the acids and bases. In the vol. for 1798, we find him fixing the relation between the metals and some of the acids, but always on the same plan.

At page xv. in the preface to the vol. for 1800, he says, "To follow an author step by step, in a path trodden by him alone, and to judge him with fairness, is not in the power of every one, still less can it be done by merely reading through his book."

Page xxiii. Again, "Whoever looks on the remarkable order, which reigns in the quantitative proportions, by which every kind of substance has a peculiar quantitative character with respect to another, as a mere play of figures, or as a mere accident, would only show his complete ignorance of the whole structure of stœchiometry, but would be indemnified for it by a still greater degree of philosophical faith; for it requires much more credulity to believe in so many accidents, than is needed to perceive that the Lord of nature has not only qualitatively but quantitatively endued it with the most wonderful order, both in great things and in small."

Another extract from the same, page 206, "In the simpler

affinities every kind of neutralizable substance has its own quantitative law of affinity, because the amount of affinities among the alkalies may be expressed by the mass, that of the acids by the substratum (that is, the body of which the oxygen of the acid is an oxide); but this is not found to be the case either with the metallic or nonmetallic combustible elements."

What then did Richter attain to is the question to be now answered. In the extract from the preface he raises the study of atomic chemistry to a science, and gives it a name. This is itself no small honour. The chemists before him had certainly not been gifted with such a clear appreciation of the importance of the study. We find that Richter has made it the leading object of his life to elucidate the laws of combination; as a young beginner, making it a subject of his inaugural dissertation, and looking forward to the time when he might have opportunity to prosecute his investigations. The word stœchiometry is preserved in Germany, with us it is too abstract for daily use.

The first definition of stœchiometry has appended to it six *experiences* (erfahrung), most of them with corollaries (zusatz). The reading becomes, therefore, exceedingly cumbrous, the words are marvellously multiplied, pure abstraction is aimed at in every step with painful strains, as it would appear, or perhaps only caused by a mathematical habit of mind too exclusively followed. In this way the few truths that we still hold to, and which are contained in the book, are so ornamented and overdressed as to have been to most persons entirely hidden under the richness of the elaboration.

He expresses his belief that the smallest portions of a body are of the same composition as the largest. He says that the affinity exists in every particle. Then adds that every piece must have the same composition. His own words are very cumbrous, but this meaning is distinctly there. This was the illustration which Dalton afterwards used on the same subject, but it was expressed in clearer words, and still earlier, by Higgins. This idea leads directly to the atomic theory and theory of equivalents. Here it is not followed out.

The sixth *experience* of first definition gives the theory of reciprocal saturation, when double decomposition takes place in solutions. This is the discovery which has been attributed to Wenzel. Let us translate his formulas into the present symbols by an example:—

$$AgO\ NO_5—NO_5,\ NO_5+KO\ SO_3—SO_3,\ SO_3$$
$$=NO_5+KO\ SO_3—SO_3+SO_3+AgO\ NO_5—NO_5.$$

He says the products of neutral salts are nearly without exception neutral, but nevertheless sees enough to form a law. Wenzel, with similar results, had not seen a law.

He endeavours to shew the relative amount of force exerted by different substances when decomposition takes place, but he gets no farther than the fact that certain forces are equal, some must be greater, and others must be less. In this district of inquiry, an example of which may be found in Theorem I., what appears to be the enunciation of an important law, frequently turns out to be the mere expression of a common-place, giving no information to the chemist. Such laws being in a certain sense universal, they are now left out of chemical works, as the mind can readily draw the conclusion for itself, if the opportunity offers.

He then shews the method of obtaining the proportion of the elements in a compound. This had been pursued with great care by Wenzel.

The great aim of Richter is not perceived in reciprocal proportion, but in the attempt to make the combining numbers of all bodies a series in arithmetical progression, and so to bring number, quantity, and order into the arrangement of the elements. In the series which he has formed, I think we may say that he has failed to prove his point. The numbers he had were too few, and the mode of obtaining the order is by no means satisfactory. There is, however, a great probability recognised by most chemists of the existence of an order in which the elements are related to each other. If this order should ever be found to be similar to that which Richter has indicated, we must do the greatest honour to his genius, although we cannot even now, when it stands before us, say that it is a discovery, or that it has any value at all.

The discovery of reciprocal proportion is given by no one before Richter as far as I know, but he himself does not speak of it as a discovery, but as a well-known fact, with which he was familiar before he wrote his inaugural dissertation. We find in the preface that it was well known that neutral salts gave neutral results on decomposition; this Richter has put formally amongst the laws of stœchiometry, and given it rank amongst chemical truths. He deduced from it, as he himself says, that there must be "distinct proportionate quantities amongst the component parts of neutral salts," and he strove hard to bring all combinations under number and quantity. The knowledge of this fact seems to have first set in motion his stœchiometry; instead then of being the point which he gained, it is the point from which he starts, according to his own account. He does not, however, seem to have seen the reason for it, nor its general bearing in chemistry, otherwise it could not have been left for Fischer to shew that the combining number of an element would fit its combination with every other element.

The mode in which he obtains the relation of the combining weights of the earths to each other is remarkably self-delusive, but at the same time exceedingly ingenious. They are given at length, so that every one may compare for himself.

He endeavours to find out a similar relation between the atomic weights of the alkalies, and readily does so. He is led away by the numbers observed to mistake them for representations of actual force, and so calculates in a relative and abstract way the force needed for decomposition. This is very characteristic of him, but unfortunately he has gone on a wrong assumption.

He has evidently been a man of great quickness, at the same time apparently of haste; he has enunciated the most beautiful truths, and left them untouched for worthless speculations, which seemed to need more ingenuity, almost leaving us to doubt how far he understood his own writings. We must, however, give him the honour of understanding what he wrote, smaller honour we can give no man. Still it is perfectly clear that if his theory were as fully developed in his mind as we with our superior opportunity can now see to be deducible from his words, men would have understood him, and the process would have been continued, but neither did he make any advance, nor did he teach others clearly, although the young Berzelius was much excited to curiosity.

It certainly is difficult to tell how discoveries grow, often impossible to tell who is the discoverer; but this we may consider a fair rule, not always easily applied, it is to be confessed, that he is a discoverer who sees distinctly the full bearing of his discoveries; when this does not happen there is a difficulty in giving that man the place due to him. It is clear that Richter, like some others already mentioned, had fundamental principles which would have led him to the atomic theory; but he has evidently been led by foregone conclusions, and the law of planetary distances has been floating in his mind and misleading him, when seeking for the differences in the combining weights of bodies.

The discovery of reciprocating proportion was a very important and memorable one, although the scientific world did not recognise it, another among the many proofs that scientific men are subject to the same bigotted attachment to the laws they have learned, as that class of men hitherto most blamed for bigotry, nor is there any bigotry more engrossing than that which appears to the possessors to be upheld by experimental proof. Who discovered this important fact, it is still left unascertained: as the expression of a law it is Richter's, but as a fact regarding neutral salts the author appears not to be known.

Among the many disputes on this point it is rather surprising that people should speak without reading the authors they discuss. The supporters of Wenzel have not read him, the supporters of Richter have overlooked his own writings and his own confession, as it appears to me in the preface. But as I give the words every one may judge for himself.

In proceeding with his inquiry one cannot but admire the energy and activity of Richter's mind, and his enthusiastic desire to prove the beauty of the arrangement of creation; it is clear that he lost his way, and spent the greatest part of his energy on a subject which could not with his data lead to great results, and which even now gives us no help, and which was not the next step wanted in chemistry. The science was straining after definite laws, it had none; Richter, with his one great law, might have done wonders, had he only seen its value; he might have found on examination that it was, properly speaking, an inference from another much more general law, and would have then expressed himself in universal terms. But Higgins had expressed himself much more clearly as to combination before him, as already shewn, and only failed because he had not seen it to be a general law.

Richter attempted to give the proportion of the acids to bases as an expression of affinity, but this had been already attempted by Kirwan, and was shewn to be unsuccessful.

As a general summary of Richter's most important work, we may say, he found that there was a certain quantitative relation between all bodies, and he made out the laws so far, that when he knew the quantitative analysis of a salt, he could tell its quantitative decomposition with another, but he never saw it with sufficient clearness to be able to express the combining quantities each by its own distinct number, nor does he appear to have ever proceeded far enough to be able to assign a cause for the phenomenon, or to connect it with any fundamental idea.

He was the founder of the systematic study of stœchiometry, he was an illustrator of one of its important laws, and a defender of regularity in nature. His scientific life was laborious, his love of science sincere, and in all respects he seems to have been a man of high character. After reading his works, and coming occasionally on a sentence which makes us for the moment believe that he has discovered a greater law than we can give to him, and finding that during his whole life he was just on the point of the present atomic laws, one feels that he was perhaps the only man that deserved to discover them, having given himself up entirely to that purpose. It is with regret, therefore, that I leave him also, another combatant who died before the victory.

It has been said that Dalton had read Richter, and had never acknowledged his claims. It is a melancholy thing to see men of talent and learning so readily distrusting their own class, as if dishonesty were so common. I might say the same of Richter, that for more than ten years he continued to publish on stoechiometry, and never once mentioned Higgins, but his whole works shew that he did not see Higgins's writings, or he would have probably got less involved than he did. We learn from Dr. Henry that Dalton had seen Richter's results on reciprocal proportion,* and had received assistance from them, but although they may have assisted him in proving his laws, Richter could never have given him

* Dr. Thomson had said the contrary; but let us take Dr. Henry's information, as being an intimate friend. Dalton could not have seen Richter's whole works, but probably an account of them. They are scarce in England. It cost me a good deal of trouble to get one, even in Germany. The British Museum does not possess a complete copy. Dalton certainly had not read Higgins; and although Dr. William Henry had a copy, we may conclude from his son's work that he had not seen in it the Atomic Theory, as he seems not to have thought it necessary to mention its existence to Dalton.

fundamental ideas. These are much wanted in Richter's chemistry. Richter's cotemporaries did not obtain the atomic theory, although some were students of his work. Berzelius himself did not obtain the atomic theory from Richter, although the most illustrious of the students of Richter's books. Dalton then could not have obtained it, and the direction he takes is perfectly different, the road he went quite clear, and the result he came to entirely distinct from that aimed at by Richter.

In such early days it required a mind of a high order to see as Richter did into the great necessity of permanent laws, and the great structure he raised to make the inquiry shews us that he saw its importance. Had chemists been accustomed to study the works of their own class, such books as his would have rapidly produced results, but the history of the matter speaks ill for the apathy of the men of even that period, and well for his untiring energy and devotion.

CHAPTER X.

FISCHER, BERTHOLLET, PROUST, &c.

When Richter had illustrated the action of neutral salts, and had seen somewhat dimly reciprocal proportion, Fischer (Ernst Gottfried) saw it in a much clearer light, and put it in a more practical form. He took Richter's analyses, and shewed that a constant quantity of one base would unite to a constant quantity of an acid, and that the numbers in all would be reciprocal. Such, one would call the true discovery of reciprocal proportion, were it not, that in this case, the intellectual labour required does not seem great.

Fischer's table is as follows :—

BASES.		ACIDS.	
Alumina	525	427	Fluoric.
Magnesia	615	577	Carbonic.
Ammonium	672	706	Sebacic.
Lime	793	712	Muriatic.
Soda	859	755	Oxalic.
Strontian	1329	979	Phosphoric.
Potash	1605	988	Formic.
Baryta	2222	1000	Sulphuric.
		1209	Succinic.
		1405	Nitric.
		1480	Acetic.
		1583	Citric.
		1694	Tartaric.*

It is explained so; "When any substance is taken from one of the two columns, *ex. gr.*, potash from the first column, where the number 1605 stands, then every number of the second column shews how much of that substance is needed to neutralize 1605 of potash." The same thing may be said

* Schweigger's *Stœchiometrische Reihen*. Page 45.

if we begin with the second column. In other words, the numbers attached to the bases indicate the amount which unites to the acid numbers (or proportions represented by them).

But we cannot say more of Fischer than that he made plainer Richter's law.

In 1803, immediately after the last of Richter's periodical publications, but ten years after the publication of *Pure Stœchiometry*, appeared the *Essai de Statique Chymique* of Berthollet.*

There is no stronger proof of the want of influence of all preceding inquirers on this subject than the existence at this period of Berthollet's essay, and the effects it had on the followers of the science, who could neither have understood nor believed the earlier, when they listened to the later with so much attention.

In Berthollet's *Essai*, we find the following sentences.

Chapter II. 13. "The chemical action of different substances is excited, not only in the ratio of their affinity, but also in the ratio of their quantity; one immediate consequence is, that chemical action diminishes in proportion as saturation advances.

14. "It also follows, from this law, that a substance which is in solution in a quantity of liquid greater than is necessary, is retained therein by a more powerful action, and that, on the contrary, the superfluous quantity of the liquid is subjected more feebly to the affinity of the dissolved substance than is required for solution. It will be seen, therefore, that the general law which I have announced is, in this instance, only modified by the circumstance which limits the quantity of liquid that can exert its action simultaneously."

Then we have this extraordinary sentence, proving that so far from the subject of the atomic theory being clear, even the doctrine of proportion had not made its way.

* The English translation by B. Lambert, 1804, is here quoted.

39. "Some chemists, influenced by having found determinate proportions, in several combinations, have frequently considered it as a general law that combinations should be formed in invariable proportions; so that, according to them, when a neutral salt acquires an excess of acid or alkali, the homogeneous substance resulting from it is a solution of the neutral salt in a portion of the free acid or alkali."

"This is a hypothesis which has no foundation, but a distinction between solution and combination."

42. "It follows, from what has been said above, that the most powerful, as well as the weakest chemical action, is exerted in the ratio of the reciprocal affinity of the substances, and of the quantities within the sphere of activity; that the action diminishes in proportion to the saturation, and that there is *no point* at which it determines the proportions, but that the limits of these proportions in the combinations which it forms, and those of its power, are to be sought for in the forces which are opposed to it. Finally, two effects of chemical action must be distinguished, that by which it produces a reciprocal saturation, and that which causes a change in the constitution."

47. "Hence it follows, that in the comparison of the acids, the first object which will fix the attention is the power with which they can exercise the acidity which forms their distinguishing character. Now this power is estimated by the quantity of each of the acids which is required to produce the same effect, that is to say, to saturate a given quantity of the same alkali."

Beautiful and ingenious as Berthollet's investigations into affinity are, he, too, missed the line of thought which was to produce the greatest discovery known to chemistry, but it was not carelessly passed over. His inquiries had led him into some very interesting qualities of bodies, the power of quantity to overpower feebler quantities in certain cases, and the capacity of bodies to decompose others according to the nature of the compound formed. For example, a salt, say an oxalate, decomposes a salt of lime, not because the oxalic acid has such a powerful affinity for lime, but because the oxalate of lime being insoluble falls, following the physical qualities of the compound; oxalic acid too decomposes common salt, not from a greater affinity, but because muriatic acid being volatile, has a way of escape, whilst oxalic acid has not, and so on.

These and similar reasonings took up such a large portion of his field of vision, that he denied, as we have seen, definite proportions, and, of course, atomic theories and equivalents could never be received from Berthollet. He denied definite proportions in the present sense, but, as Berzelius says, allowed them, although within certain limits. This would imply proportions of a rather indefinite kind. In the "Mémoires de la Société d'Arcueil," he even gives laborious analyses tending to fix the proportions of the elements in certain bodies. This was chiefly from the point of view that distinct bodies had determinate and constant composition, which, of course, it would have been too late to deny.

Berthollet did, nevertheless, put a drag on the inquiry into the general question of proportion, and from his superior position commanded great influence. It is in this state we find the science, then, when Dalton came to it. It may be said that the state of the science comprehended all that was published, and this is in a larger sense true, and constitutes the base of a final opinion of a discoverer's merits; but the state of what seemed to be the well-grounded science, as far as the leading men held it, was rather that of Berthollet than any others, the leading innovators were obscure, and never indeed became magnates in the scientific world. These latter commanded to all appearance in the capital, whilst the real power was getting prepared in the provinces.

Berthollet's essays must always stand as the greatest proofs of the reality of Dalton's achievements. Whatever men may have built before, this is the existing ground on which Dalton had to raise his edifice.

To some extent this may be reckoned questionable, as the discussion as to the definite proportions of compounds was going on with Proust, a strong supporter of this doctrine, but this exception only shows that the differences of opinion were raised, not on the great theory, the history of which we are discussing, but on the simplest preliminary facts; the laws themselves did not appear in the discussion from the cause already given, viz., virtual non-existence, except in a state of possible evolution from laws which themselves were not fully proved.

I shall now give those sentences from Proust's writings which seem most nearly to affect our subject.

* "As to those which have been announced by Thenard, I will not contest results obtained by a chemist who knows how to operate with that exactitude which characterizes a consummate worker. I will say, however, without attaching any importance to my opinion, that in considering this almost general law of nature which offers us everywhere only one or two terms of oxidation of metals, and from which, in our artificial imitations, we cannot free ourselves, I fear that the six terms which he recognises are not all sanctioned by nature.

"If by the assistance of a high temperature we lower the weight of an oxide which is at its maximum, and which does not happen to be volatile, or if by a high continued heat we elevate a metal to its highest oxidation, are we to believe that all the ascending or descending terms of oxidation, which may be inserted between the extremes, are to be taken as so many different terms of oxidation? Certainly not. I do not recognise in that the ordinary course of nature, and I venture to believe that in similar cases we only make mixtures in all possible proportions of the oxide at minimum with the oxide at maximum.

* Journal de Physique. Tom. 55, page 331. 1802.

"I will relate a few facts very suitable for rendering this view plausible, if they do not confirm it. In analyzing, as I have had occasion to do, some of the oxides of all degrees which Rinman has announced in calcining steel, iron, and cast iron, I have met with only two known oxides of this metal, mixed in different proportions; the ores yielded me only an equal mixture of black and red oxide.

"If we examine the greenish oxides which lead and bismuth give at the commencement of calcination, and those of tin, copper, &c., we only find an oxide at the maximum which envelopes different portions of the metal. If now we measure these degrees of oxidation by the nitrous gas which they give, we shall be led to believe that the metals oxidize themselves in all doses. But, no. There are unions of oxygen like those of sulphur, acids, &c. Election (elective affinity) and proportion are two poles, round which invariably move all the systems of true combination, whether in nature, or in the hands of the chemist. In a word, oxygen is not one of those bodies which can be mixed; when it combines, it subjects itself to certain proportions, and these proportions are what we have now to study."

1806. Vol. lxiii., page 367.

Defending Klaproth, he says, "A combination, according to our principles, Klaproth would tell you, is a sulphuret of silver, of antimony, of mercury, of copper; it is also a metallic oxide; it is a combustible body acidified; it is a privileged production to which nature assigns fixed proportions; it is an existence which nature has never created even in the hands of man, except with the balance in her hand *pondere et mensura*. Know then, he would add, that the characters of true combination are invariable as the proportion of their elements. From one pole to the other they are found the same under these two aspects, their physiognomy only may vary according to their mode of aggregation, but never their properties. No differences have been observed between the oxides of iron of the south and those of the north. The cinnabar of Japan has the same proportions as that of Almaden. Silver is neither oxidated nor muriated differently in the muriate of Peru from the muriate of Siberia. In no part of the known world will you see two muriates of soda, two muriates of ammonia, two saltpetres, two sulphates of lime or of potash, soda, magnesia, or baryta, which are different; in fine, it is with one measure that all the combinations of the globe have been formed."

Page 370. "Nature has imposed certain laws of proportion in relation to those unions which we have come to call combinations."

Vol. lxiii., page 439.

"None of the researches which have been undertaken hitherto to assist the hypothesis of variable oxidations, even among the class of nonmetallic combustibles, have been able to discover above one or two of each; and each of these once oxidized is equally a product, the characters of which are invariable, preserving its properties with firmness on all occasions where they can be shown, whether free or in a state of combination. To this height are we now arrived in this branch of natural science."

Page 466. "There is nothing whatever in opposition to our extension of the same principles by regarding the solutions of sulphur, phosphorus, carbon, arsenic, zinc, &c., in hydrogen, not as simple solutions, without measure, in unfixed proportions, but as proportional combinations, as hydrurets of sulphur, phosphorus, &c., which the excess of the solvent may take into solution."

"Sur les Sulfures Métalliques."

Jour. de Ph., vol. lix., page 261, year 1804.

The following is a portion of the controversy between Berthollet and Proust. It begins with quotations from Berthollet.

"He (Proust) believes that there is attached to antimony

a dose of sulphur invariably fixed by nature, and that it is not in the power of man to increase or diminish it. He fixes this at 25 per cent."

Proust continues, "It is not I, but nature, or whatever power you choose, which places a barrier between it and all the efforts of every chemist who will attempt to make sulphuret of antimony above or below this proportion. *I have assigned no law to my discovery;* I have verified it only; I have followed the precept which Berthollet himself has traced in his profound work; when, says he, one substance combines with another, it is necessary to determine the proportions, and to examine the properties, &c. Such, in fact, has been the constant object of chemists, from the moment that they recognised that this determination was one of the most important bases of the history of combinations and of the science of analysis. Nobody can believe that nature will abandon her compounds to the chance of those variable proportions which Berthollet has chosen as the foundation of his system. But it is not the less true, that in proportion as the horizon of sulphurets extends, we do not see that the new facts which every day accumulates are of a nature to strengthen it."

Berthollet against Proust. "He has combined the oxide of antimony with different proportions of sulphur, and has obtained mixtures which may be represented by this formula; oxide $+1+2+3$, &c., of the sulphuret of antimony: has he not obtained there veritable combinations?" Page 262.

Proust: "To this, I shall reply, that solutions which have commenced, or which have not attained the term of saturation of which we consider them capable, ought to be viewed differently from combinations which are completed; but to explain myself, I illustrated those solutions in the same way I would do those of sugar in water, that is, as water $+1+2+3$ of sugar. I do not see that we can form more distinct ideas of the solutions of the sulphuret of antimony in its oxide. All chemists have hitherto believed that these glasses, livers, and crocuses, were oxides which had been sulphuretted. The object of my work has been to show the fallacy of this, and that we must give up these sulphuretted oxides, which we admit, without proof, and receive in their place a species of combination, new, without doubt, but well demonstrated. Certainly this combination is in opposition to the ideas of Berthollet; he wishes to place them in the family of sulphuretted oxides, but it is no less certain, that those which I have announced do exist, and that they have this advantage over sulphuretted oxides, the existence of which is now terminated, that they afford the most natural solution of those thousand and one problems in antimony, the ridiculous nomenclature of which has shown the confusion of our ideas, and covered with obscurity the history of that metal."

Page 264. "To a ℔. of potash you add an ounce of arsenic; it is not saturated, you add two, you add three, it is not yet saturated, and so with more; but in waiting to discover the point of saturation, I repeat to them; your arsenical potashes are nothing at present but potash, plus one, two, or three of arsenic, but we have not time to prove that this combination will obey, as it no doubt will, the law of proportion, and shall not press you to decide on it. These are results so variable that they destroy your laws of proportions, and render your apothegms illusory. Berthollet is too just not to agree that the series of numbers by which I have sought to represent the solutions of the sulphuret of antimony in its oxide, has not the least relation to that which I have hitherto called proportion in combinations."

Journal de Physique, 1805, vol. lix., page 321.

"Sur les Oxidations Métalliques."

"I ought to explain, says Berthollet, that the proportions of oxygen may vary progressively after the limit or the combination becomes possible, until it attains the last degree; and when this does not take place, it is because the conditions which I have indicated become an obstacle to this progressive action."

"A little before this are to be found the facts on which this illustrious chemist establishes the theory of progressive oxidations, which he opposes to that which I have given out on different occasions, and of which the base is, that the combustible bodies are arrested at fixed terms of oxidation, in the same way as we see to be the case with sulphur, phosphorus, carbon, azote, and the greater part of the metals."

Page 328. After shewing that the fine lead powder got by shaking the metal in a bottle is a mixture of metal and oxide of lead, and not a low and varying degree of oxidation, he adds;

"But is each of these molecules, one might say, suddenly at one leap elevated from 0 to 9, to 12, to 25 per cent? And is it possible that they do not pass successively all the ascending terms which the imagination can conceive possible between the two extremes! I reply, that it is impossible in the actual state of things to say if the oxidation follows or does not follow this progression, because in the calcination of a metal, of lead for example, the senses are not struck with any phenomenon which can guide the judgment in the choice between two opinions; but although we do not see intuitively that which occurs actually in calcination, we are not hindered from judging clearly by the aid of numerous analogies which the field of combination offers."

Page 330. "Let us mix the green sulphate of iron with the red, each base will hold its own amount of oxygen, and there will be no conciliation between the two, no division which will bring forward to us those intermediate oxidations which the mind would desire to discover. Is it a piece of iron which we throw into the red sulphate? We see the base of the salt descend to 28, not by a retrograde march which arrests each of these particles at 47, at 46, at 45, &c. of oxidation, but by the instantaneous lowering of each from the limit (or term),

48 to 28, and this is well confirmed by analysis, because we discover in the solution nothing but red molecules mixed with green ones."

Page 334. " I will say then of the sulphurets as I said of the oxides, there are only two of them."

Speaking of the oxide of copper:

Page 351. " If we establish the calculation on this basis, we find that a quintal of yellow oxide is composed of 86 of copper and 14 of oxygen, whilst the black oxide to which we wish to compare it, contains only 80 of metal and 20 of oxygen; or in other words, if copper condenses 25 per cent. of oxygen to raise it to its maximum, it condenses only 16.3 to raise it to its minimum. Here then, we find, as in all other combinations of oxygen, new reasons for recognising this law of nature, which subjects the metals and combustibles to those proportions from which we cannot separate them, however various may be the circumstances under which they have united."

It really is a melancholy thing to read these papers of Proust. He had advanced by the most careful steps to the conclusion, that all combinations were made in proportions defined by some law of nature, that they were weighed and measured before they were united, and yet failed to see a law. Richter used the words from the Septuagint, God has made all things by measure, number, and weight; and Proust uses a similar phrase with his "pondere et mensura," from the Vulgate, leaving out the word number, which he did not sufficiently see. He saw, with great clearness, that without such constant proportion the products of nature would lose their stability, and the characters of bodies could not be depended on for permanence. We have here no difficulty in judging how much he did, and how much he left undone; how far his own mind was advanced, and how it had merely speculated. He tells us all distinctly. When he uses +1+2+3 of proportions, he tells us it is merely for illustration, he did not mean it to indicate the order of combination, he had in fact made no theory, at least found no law on the subject, although he clearly saw that it must be owing to some law of nature. He fought for constant proportions in combinations, and fought well, but he had no idea of a constant quantity of oxygen found uniting with a constant quantity of every metal, and making higher oxides by steps always of an equal altitude, although he proved that the rise was not that of an inclined plane, but by "fixed terms." And yet it follows as a consequence, so closely in fact does it follow, that we must put ourselves in the position of the early chemists of the century well to understand the difference. When we have taken that position, we then see how thin was the veil, although utterly impenetrable, that divided his opinion from the present, and prevented the acute, active, and logical mind of Proust from attaining to the great discovery. His determinate proportions are given as remarkable facts, in connection with which he confessed to perceive no law.

Had such men studied Higgins, we should probably never have heard of this controversy, but he was not studied. We may therefore learn not so readily to blame a man for want of honesty, when he publishes for a discovery what has been known before. The most indefatigable workers of the period had neither read Higgins nor Richter. Besides, scientific men like other men are led by fashion, the follies of some men become great discoveries for a while, and the wisdom of men comparatively obscure, such as the two mentioned, is neglected or sneered at. Yet the whole body endeavours to acknowledge facts only.

As a specimen of the chemical books of the time, let us take one published in the same year as Dalton's work on the atomic theory, although after Dr. Thomson had made it public, "A Course of Theoretical Chemistry, by Friedrich Stromeyer."*

* Grundriss Theoretischer Chemie zum behuf Seiner Vorlesungen entworfen von D. Friedrich Stromeyer. Göttingen, 1808.

He says, p. 66, § 36, "The affinity of a substance towards another is always in proportion to its chemical mass, which it brings with it for combination."

Page 68. "Affinity is consequently in no way an elective attraction, in the sense in which Bergman and his followers have taught it."

Page 80, § 59. "By means of affinity alone two substances may combine with each other in every quantitative proportion. Although this asserts quite the opposite of Bergman's theory, and even appears to contradict experience, it is nevertheless a perfectly natural consequence of the above.

"For suppose A and B could combine in the proportions of 5 : 7, and not as 10 : 7, if we put 7 parts of B with 10 of A, then the very same thing would happen as when we put 7 parts of B with 5 of A; that is, the 7 parts of B would combine with 5 parts of A, and the other 5 parts would remain uncombined.

"If this were really possible, then it would follow clearly that 10 parts of A contained no more affinity than 5 parts, which however completely contradicts what is proved in § 36."

The idea of the present law seems to have entered into his mind as a conclusion to be avoided for its absurdity.

§ 60. "But as soon as the power of cohesion or expansion of two substances of a given mass, acting on each other, ceases to be overpowered by their affinity, an exception constantly takes place to this universal law, and a certain proportion of mixture is established between the two substances, which, on account of their affinity, can no longer be surpassed, and consequently limits their mutual affinity.

"But if there were a removal of the obstructions caused by cohesion and expansion to the affinity of two bodies, then the law would again come into action in its first unity."

"The absorption of several gases by water, the solution of salt in water, and the oxidation of metals, best establish that which has been said."

§ 61. "This general law is subject to an exception also, when there is a change of the aggregate condition during the action of two or more substances, and in this case also a combination takes place of a fixed proportion."

Here we have a clear account of the direction that Berthollet's teachings gave, and the consequences are logically deduced. Here, too, we find that fixed proportions are obtained as exceptions, but it is also seen how needful it was to have them occasionally in order to explain facts in the science.

Stromeyer's words do, in fact, represent the confused and contradictory opinions of the time, and afford us another proof that no one before Dalton had given opinions sufficiently authoritative on the atomic theory to be retained by the teachers of chemistry; and we may add also, none deemed of sufficient importance to demand at the time very serious discussion.

This want of attention, even to imperfect theories, arose mainly, I believe, from the fact, that those theories hitherto given had not had accumulative scientific proof to give them force in the world; they had organized no executive force. Chemists, generally, had not arrived as far as the inquiries already quoted.

It was intended to give a number of similar instances, to show how entirely all atomic theories and theories of definite proportions were out of the boundaries of general chemical science when Dalton published. I happened to take up Stromeyer first; many instances occur in our own country, but this will probably be found sufficient.

CHAPTER XI.

DALTON'S ATOMIC THEORY.

We now come to Dalton, whom we have left since he first thought of weighing atoms; this has been done in order that by accumulating all the materials he might have used, we may know what has been his especial work. Matters stood thus: Higgins had, in 1789, seen the fundamental principle clearly, and given it out more as a thinker than a discoverer, neglecting to generalize; that principle included simple proportion or the law of necessary definite composition and multiple proportion. Richter had systematized all the laws of combination known to him, but had not known of Higgins, although he would there have got the clue to all his strivings. He discovered one of the most important consequences of the fundamental law, in reciprocal proportion, but did not rise up to first principles. Scientific men could get no decided guide from either, and preferred to follow Berthollet, who was leading them out of the right direction, obstructing for many years the advance of chemical philosophy, or compelling others to accumulate proof until it was sufficient to overwhelm him.

The materials we have for tracing the progress of Dalton's opinions are few, but distinct and sufficient. I shall not enter into the argument of honesty, which can be thrown with greater violence at so many heads, but shall take it for granted, that all those who have had the honour of working at the laws of chemical combination with any success, have also had the advantage of an honest mind. I know of no dishonesty on any side among the principals connected with this subject, and their defenders have erred, probably, on account of proving one side better in morals than the other.

From the earliest period of his scientific life Dalton had been accustomed to think carefully on the constitution of the atmosphere; this is seen as early as 1793, in his meteorology. This subject continued to be a favourite one, and led him to gases generally. The experiments quoted at p. 43, on nitrous gas and oxygen, and those mentioned afterwards in a quotation from Dr. Thomson, shew the method by which he came to believe, and to prove experimentally, the existence of definite and constant proportion.

Here lies the difference between him and Higgins. Higgins expressed the fundamental idea as clearly as Dalton, but it was still left uncertain. Dalton proves it by experiment, draws the conclusion, and tells us the "theory" in a few lines. We have then distinctly the method by which he came to believe in definite proportion and multiple proportion. He proved them for himself, and theorized for himself. No books of any writers before him, no Proust or Berthollet controversies were so decisive as these few experiments of Dalton; not clearer than the words of Higgins, but more decisive, because the result of observation and of reasoning combined. This seems to have been his first direct entrance into the region of the atomic theory.

In reading over his earlier works, or even in reading the short account here given, we may remark with what a firm grasp he lays hold of the existence of atoms, of the idea that all matter is made up of separate ultimate particles, divisible or indivisible. We find no scientific man holding the idea with such firmness; to others it was a theory, to Dalton it was a fact, which he could not conceive otherwise. We find that even the air is represented* as an agglomeration of bodies heaped up like piles of shot. We appear to be entirely removed from the region of speculation when reading his words; although he leads us farther than the most fantastic speculator had done, the road is made so clear before us, that

* Page 47.

we find no difficulty, either physical or metaphysical. He persuades us to go, leads us and describes all to us in a few sentences, when volumes of persuasion written before had not been sufficient to induce men to turn their eyes. His descriptions are rigid as well as picturesque. Some persons would apply to them the word material, and still more the word mechanical. It was by following rigidly the *mechanical* properties of his atoms that he arrived at his results. To those who read his works, it will be clear that his mind became gradually more confirmed in this course.

In reading what he says, at p. 49, we see him plainly verging towards his theory, and also the nature of his struggle, which is in no respect similar to that of any other inquirer. He there first says, that he is inquiring into the relative weights of the ultimate particles of bodies. This idea had never suggested itself as practical to any one before Dalton, nor am I aware that it has ever been claimed. Having made some advance in this inquiry, he made it the starting point of all that he advanced on atomic chemistry and the theory of proportion.

He was not in haste to publish his theory, but told it openly to Dr. Thomson, in 1804; this, then, is the date of the complete discovery, as Dr. Thomson published an abstract of it at once. Some persons unacquainted with this have advanced an argument against him, his own book containing the subject not having been published till 1808. A well known quotation from Dr. Thomson's history says, "Mr. Dalton informed me that the atomic theory first occurred to him during his investigations of olefiant gas and carburetted hydrogen gas, at that time imperfectly understood, and the constitution of which was first fully developed by Mr. Dalton himself. It was obvious from the experiments which he made upon them, that the constituents of both were carbon and hydrogen, and nothing else. He found, further, that if we reckon the carbon in each the same, then carburetted hydrogen contains exactly twice as much hydrogen as olefiant gas does.

This determined him to state the ratios of these constituents in numbers, and to consider the olefiant gas a compound of one atom of carbon and one atom of hydrogen; and carburetted hydrogen of one atom of carbon and two atoms of hydrogen. The idea thus conceived was applied to carbonic oxide, water, ammonia, &c., and numbers were given representing the atomic weights of oxygen, azote, &c., deduced from the best analytical experiments which chemistry then possessed."

His first atomic weights, already given,* were published in 1803; he did not publish his "New System" till 1808. He says then, † "A pure elastic fluid is one, the constituent particles of which are all alike, or in no way distinguishable. * * * These fluids are constituted of particles possessing very diffuse atmospheres of heat, the capacity or bulk of the atmosphere being often one or two thousand times that of the particle in a liquid or solid form. Whatever, therefore, may be the shape or figure of the solid atom abstractedly, when surrounded by such an atom it must be globular; but as all the globules in any small given volume are subject to the same pressure, they must be equal in bulk, and will, therefore, be arranged in horizontal strata like a pile of shot."

The chapter "On Chemical Synthesis" gives his theory. He there says, "When any body exists in the elastic state, its ultimate particles are separated from each other to a much greater distance than in any other state; each particle occupies the centre of a comparatively large sphere, and supports its dignity by keeping all the rest, which, by their gravity, or otherwise, are disposed to encroach on it, at a respectful distance. When we attempt to conceive the number of particles in an atmosphere, it is somewhat like attempting to conceive the number of stars in the universe; we are confounded with the thought. But if we limit the subject, by taking a given volume of any gas, we seem persuaded that, let the divisions be ever so minute, the number of particles must be finite; just

* Page 49. † "A New System of Chemical Philosophy." Part I., p. 145.

as in a given space of the universe, the number of stars and planets cannot be infinite."

"Chemical analysis and synthesis go no farther than to the separation of particles one from another, and to their reunion. No new creation or destruction of matter is within the reach of chemical agency. We might as well attempt to introduce a new planet into the solar system, or to annihilate one already in existence, as to create or destroy a particle of hydrogen. All the changes we can produce consist in separating particles that are in a state of cohesion or combination, and joining those that were previously at a distance.

"In all chemical investigations it has justly been considered an important object to ascertain the relative *weights* of the simples which constitute a compound. But unfortunately the inquiry has terminated here; whereas from the relative weights in the mass, the relative weights of the ultimate particles or atoms of the bodies might have been inferred, from which their number and weight in various other compounds would appear, in order to assist and to guide future investigations and to correct their results. Now it is one great object of this work to shew the importance and advantage of ascertaining *the relative weights of the ultimate particles, both of simple and compound bodies, the number of simple elementary particles which constitute one compound particle, and the number of less compound particles which enter into the formation of one more compound particle.*

"If there are two bodies A and B which are disposed to combine, the following is the order in which the combinations may take place, beginning with the most simple: namely,

1 atom of A+1 atom of B=1 atom of C, binary.
1 atom of A+2 atoms of B=1 atom of D, ternary.
2 atoms of A+1 atom of B=1 atom of E, ternary.
1 atom of A+3 atoms of B=1 atom of F, quaternary.
3 atoms of A+1 atom of B=1 atom of G, quaternary.
&c., &c.

"The following general rules may be adopted as guides in all our investigations respecting chemical synthesis.

"1st. When only one combination of two bodies can be obtained, it must be presumed to be a binary one, unless some cause appear to the contrary.

"2nd. When two combinations are observed, they must be presumed to be a binary and ternary.

"3rd. When three combinations are obtained, we may expect one to be a binary, and the other two ternary.

"4th. When four combinations are observed, we should expect one binary, two ternary, and one quaternary, &c.

"5th. A binary compound should always be specifically heavier than the mere mixture of its two ingredients.

"6th. A ternary compound should be specifically heavier than the mixture of a binary and a simple, which would, if combined, constitute it, &c.

"7th. The above rules and observations equally apply when two bodies such as C and D, D and E, &c., are combined."

* * * * * * *

"In the sequel the facts and experiments from which these conclusions are derived will be detailed, as well as a great variety of others, from which are inferred the constitution and weight of the ultimate particles of the principal acids, the alkalis, the earths, the metals, the metallic oxides and sulphurets, the long train of neutral salts, and in short, all the chemical compounds which have hitherto obtained a tolerably good analysis. Several of the conclusions will be supported by original experiments.

"From the novelty as well as importance of the ideas suggested in this chapter, it is deemed expedient to give plates exhibiting the mode of combination in some of the more simple cases. A specimen of these accompanies this first part. The elements or atoms of such bodies as are conceived at present to be simple, are denoted by a small circle, with some distinctive mark; and the combinations consist in the juxtaposition of two or more of these; when three or more particles of elastic fluids are combined together in one, it is to be supposed that the particles of the same kind repel each other, and therefore take their stations accordingly."

In the figures to which he refers above he has shewn to us how vividly he formed these ideas, that they were no mere fancies which had passed through his brain, but distinct impressions, ready prepared for utterance. No doubt is left upon our minds as to his opinions, which are, that every piece of matter, even the smallest, must follow the laws of the largest; that when pounds of matter unite, the atoms contained in them must unite also, until we come to the fact that only atoms can really be said to unite. Now as the conception of any fraction of an atom is a contradiction and impossible, they must constantly unite as wholes, and the proportion will be constant. If constant in the smallest quantities, then so in the largest, explaining the permanency of the constitution of bodies so much disputed, and making it a law of nature. If two compound bodies unite, the same law is followed out.

He then gives instances of combination, and adds to his explanation a plate of the "arbitrary marks or signs chosen to represent the several chemical elements or ultimate particles."

He gives twenty atomic weights and seventeen analyses of gases and acids. His atomic weights are—

Hydrogen, its rel. weight..	1	Strontites	46
Azote	5	Barytes	68
Carbone	5	Iron	38
Oxygen	7	Zinc	56
Phosphorus	9	Copper	56
Sulphur	13	Lead	95
Magnesia	20	Silver	100
Lime	23	Platina	100
Soda	28	Gold	140
Potash	42	Mercury	167

"An atom of water or steam composed of one oxygen and one hydrogen, &c. = 8," and so on with other bodies.

This is the result of enormous labour, added to that of the many gaseous analyses. Let no one say that because the atomic weights are in most cases inexact he shews want of power. We have seen in our own days how difficult it is to get an exact atomic weight, we have found that it needs the combined forces of several laboratories to settle one to satisfaction, and we must rather admire that man who approached first so near. But, although Dalton has been called a rough worker, and I am not prepared to deny it, we must remember that his analyses are not behind the time, but in advance of it in early life. At the period when he was working out this theory, the analyses of all chemists were in general only approximative. Fine analysis was only then beginning its course. But as Dalton says, "it is not necessary to insist on the accuracy of all these compounds, both in number and weight; the principle will be entered into more particularly hereafter, as far as respects the individual results."*

He also says, "it is not to be understood that all those articles marked as simple substances, are necessarily such by the theory;" *ex. gr.*, soda and potash are mentioned as compound.

In various parts of his work we learn exactly the method in which he applied his theory, and as he devoted his time to its illustration, we are not left in any doubts as to his opinions.

We have now to find exactly what new ideas he produced. We have seen in the last chapter the state of chemical opinion, the prevalence of the Berthollet philosophy, and the uncertainty hanging over the opinions opposed to his. General opinion on combination was in reality not more advanced than in the earliest days of the science. Dalton found matters in this state of confusion, and we have seen the

* Page 220.

results he arrived at, and the process. 1st. By long reflection on the constitution of bodies, especially of gases, he became convinced of the necessity for ultimate particles, divisible or not so. These particles unite together, and form of course a definite compound. If the smallest part is definite, so is the largest. This is the fundamental law of definite compounds. 2nd. Various numbers of atoms may unite—there may be one, two, three, or more—there can be no division of atoms. As large bulks are constituted just as small are, so multiple proportion becomes a law by which bodies are constituted. 3rd. Compound bodies constituted as the above unite particle to particle in a manner exactly similar to simple bodies, and so we have compound proportion, and are led to a mode of inquiry into, and a method of expressing the most complicated bodies. 4th. If we obtain the relative weights of the constituents of bodies we obtain the relative weights of the atoms, because the smallest parts must be constituted as the largest. 5th. The relative weights of the atoms become constant expressions for the proportions of combinations.

These are the fundamental principles which made chemistry a science and hold it together, and although Dalton had no direct help in discovering any of them, we have seen that Higgins had already expressed the first two. Richter and Fischer had made out numbers representing the reciprocal proportion of bodies, and although not going to first principles, and establishing no law, Fischer's numbers adopted by Richter, I believe in 1803, are really atomic weights or equivalents, although they did not see them to be such.

There existed, therefore, in the world material for completing the theory of combination, but there was no one who saw it clearly, and no one who knew both parts published. Dalton cannot be blamed for not knowing them; no one knew them. Although Dalton had to begin without their aid, the custom of the world is to give credit to him who adds to its accumulated knowledge, not to him who obtains knowledge in his closet merely. By this custom, therefore, Dalton's credit would be the whole theory, minus what Higgins and Richter had done; how much this was who will venture to declare, the two parts being the conceptions of different brains, so that Thomson, with his great quickness, could make nothing out of one, and Berzelius, with his patience, nothing out of the other; nevertheless, Dalton must not have their credit, and it was not the habit of his independent spirit to seek from others. Dr. Henry has made it clear that he had not seen Higgins's book. He no doubt saw Proust's results. They are in the library of the Manchester Society, and they are probably the most likely to have assisted him, as the analyses and reasoning are remarkably clear, but devoid, as before said, of theory; still the clearest and best were later than Dalton. This, therefore, seems to be the result, that although actually producing all the theory within himself, from the world his deserts are that he first saw the great importance of the idea of using atoms to illustrate proportion and definite constitution. He followed up the idea, and found in it a fundamental natural law, as it appears hitherto. He saw the use and importance of multiple proportion, or the adding of atom by atom in twos or in threes, and he proceeded to investigate nature under this impression. He proved that bodies followed laws, such as fitted his hypothesis, which was thenceforth taken into the province of scientific theory. To perform the above it was needful to grasp the idea more firmly than it had been done, to work laboriously, and to decide convincingly. This Dalton did. He then extended this from simple to complex combinations, and gave the first idea as well as proof of compound proportion, laid down the laws orderly into a system, and accompanied the whole by abundant and laborious proofs. He gave the first idea of atomic weights. Under this head came Richter and Fischer's numbers. Richter grappling with those numbers never could obtain a rational theory from the phenomena. Dalton's plan explains these numbers with the greatest ease, and looks on such as a necessity of the fundamental law, instead of the beginning of the inquiry as it was to them.

It seems to me, then, that what happened historically, happened also intellectually. Dalton had included his predecessors in his more extensive system. He had gone to the summit of the hill, and when coming down, found proofs that they had been making good progress upwards. Higgins had gone at once to the top, as it appears to me, but took no heed to make the needful observations when he was up, or he found the prospect entirely obscured. We are compelled to put reciprocal proportion in a secondary position, as it seems to me it cannot be called a law, but one of the consequences of a law; and the evidence brought to support it, otherwise than empirically, presupposes some of the principles on which the general laws depend.

It was by a careful mechanical juxta-position of parts that Dalton arrived at his idea, it is eminently mechanical, and it is remarkable that all progressive views on that subject have been so. *He introduced proportional weight into the theory, and found it to agree with facts.* His is, therefore, the *quantitative* atomic theory. In this complete form no one seeks to take from him the honour. The total is so entirely his, that the disputed parts can be held only as a fealty.*

Although Dalton rigidly held to the idea of atoms, he by no means supposed that we had attained the indivisible atom

* Dr. Schweigger, in his pamphlet, objects that Dalton's theory was not conceived in the spirit of the ancient theory, because he allowed some atoms to be small and some large. Strange this. The reason that the ancient theory is insufficient, is simply because it was not conceived in the spirit of Dalton's. 2nd. Most of the ancients allowed greater and smaller atoms, and various shapes. 3rd. Dalton formed his theory in the belief that atoms were of one size, but afterwards saw reason to change his opinion. Dr. Schweigger, therefore, has forgotten the opinion of Dalton and the ancients, as well as Richter's preface, where he calls the permanent neutrality of neutral salts after neutral decomposition, a well known fact. He sneers at the atomic theory, thinking that by putting it down, Dalton will fall; I don't agree with him there, but there is time enough to wait.

in our elements; at least, he expressly reserved this point. What he speaks of is simply the ultimate particle that seems to act in our chemical processes.

Dalton used atom and particle. Many have objected to both, but they are words which really involve less theory, and are more generally applicable than any yet obtained, except, perhaps, combining proportion, which is too long. Equivalent, Wollaston's word, is itself too long, limited in its meaning at best, and at times either meaningless or incorrect. Whenever we conceive of combining proportion, we reduce the quantity to the smallest conceivable, every portion becomes a unit, and each unit is undivided; it is a particle for the moment, and an atom as far as the effect we study is concerned. Having, after much labour, attained this mode of thinking, we can now scarcely think of compounds otherwise, nor indeed has any other method suggested itself as better, although investigation is promising advance in our knowledge, and will probably some day astound us by its results.

The consequences of Dalton's laws gradually shewed themselves to be, that there was now one great law or theory in chemistry, so that it was for the first time fit to be called a science. Heretofore it was a series of separate facts, and even now we may say that it is more a branch of natural history than of exact science, but as a science it is preserved by this fundamental law and its branches, and by this only. It is stretching itself out in many directions, and its future will undoubtedly be still more brilliant; at present, although it has planted its standard in numerous spots, it has fixed a government on few, but still issues the central law unchanged, although with explanatory extensions. The history of the atomic theory, since Dalton's time, is contained in Dr. Daubeny's work, and on that it is not my desire to encroach, nor could I hope to equal it.

A very little need be said as to the consequences of the promulgation of Dalton's views. They were explained in

1804 to Dr. Thomas Thomson, the very learned professor of chemistry in the University of Glasgow, and by him they were inserted in his system of chemistry. But it was not a truth that could electrify the world, it was on a subject on which few thought, one of which many said, "what is the use of it?" that miserable question which occurs to men to whom the revelation of God's truth is of no interest, unless an immediate advantage is promised. Dr. Thomson (Nicholson's Journal, Vol. 21, p. 87) says even four years after, "This curious theory, which promises to throw an unexpected light on the obscurest parts of chemistry, belongs to Mr. Dalton;" shewing that even then it required to be taught even to chemists. We find in reality that for years afterwards it was to most chemists a mere speculation.

In a paper by Wollaston in the same volume, p. 164, he says, "Dr. Thomson has remarked that oxalic acid unites to strontian as well as to potash in two different proportions, and that the quantity of acid combined with each of these bases in their superoxalates is just double of that which is saturated by the same quantity of base in their neutral compounds. As I had observed the same law to prevail in various other instances of superacid and subacid salts, I thought it not unlikely that this law might obtain generally in such compounds, and it was my design to have pursued the subject with the hope of discovering the cause, to which so regular a relation might be ascribed.

"But since the publication of Mr. Dalton's theory of chemical combination, as explained and illustrated by Dr. Thomson, the inquiry which I had designed appears to be superfluous, as all the facts that I had observed are but particular instances of the more general observations of Mr. Dalton, that in all cases the simple elements of bodies are disposed to unite atom to atom singly, or if either is in excess, it exceeds by a ratio to be expressed by some simple multiple of the number of its atoms."

It is plain that Wollaston, who was behind in no theories, was unable to obtain from the knowledge of the times an explanation of these phenomena, even when he saw the difficulty clearly; and it is plain also, that although it was said that he would have discovered it had Dalton failed, that he did not discover it even when he had in his hands as many facts as Dalton had, and a greater power of accurate workmanship. The genius was wanting, the acuteness of Wollaston and of Proust could not penetrate, where the simplicity of Dalton was at home.

In Sir Humphrey Davy's Bakerian lecture for 1809, in giving some analyses, he says, "the same proportions would follow from an application of Mr. Dalton's ingenious supposition," but even he, with a mind much more capable, as we might suppose, of delighting in grand general laws, such as from their brilliancy come upon us like the finest poetry, even then saw little in the new theory, and said little upon it. When it began to be established he was inclined to prove that Dalton was not the first discoverer, although when he delivered his discourse on giving Dalton the gold medal of the Royal Society he had arrived at a clearer view of the subject. The same slow vision was, as might be more readily conceived, the case with Berzelius, who took it gradually up as it came from his own analyses, growing during his whole life into more and more absolute certainty. But Dalton had given a table of atomic weights in 1803, and for years these philosophers plodded after him with numberless proofs, Berzelius surpassing all others in accumulating and arranging scientific wealth.

It will not be pleasant to review the delinquencies of our great men, nor the slowness of their conceptions, their desire to limit the law, and to cramp it down to the bounds of their own knowledge, denying its power to explain more than they had seen. Scepticism has its work to do in the world, and credulity has had many victims, but it may well be said that the amount of knowledge to be gained is so great that our capacities for belief must enlarge themselves rather than diminish, and we shall be left behind with a small and inadequate supply of intellectual food, if we refuse to take it until we have extracted its one-tenth per cent. of questionable adulteration.

The law has established itself; it is true. Our knowledge of its ramifications will increase. If it had not great ramifications, we might question its own intrinsic greatness. Isomorphism and isomerism are two of these, beautiful in every respect, thoroughly beautiful. But they are no contradictions, the number of provinces do not prove the smallness but the greatness of a kingdom. Why then should they even be mentioned as modifications, they may be said even to be necessary results. Allotropy is itself a curious and beautiful fact, and one that we may readily suppose will widen itself out, perhaps even till alchemy itself shall cease to be wonderful; but it does not disprove the atomic combinations.

We may call the ultimate particles which practically unite *equivalents*, as Wollaston did, but we don't alter the fact, we only chronicle our hesitation, and substitute a name which cannot be final, as it represents only a temporary theory; it only says that the quantities are equivalent to each other, but refuses to decide what these quantities are. We may call them volumes, like Berzelius, but we find then that we go on a wrong hypothesis, as the atomic volumes are not the same as their weights in every condition, whatever might be the case if all were gaseous.

We have, in fact, found no name representing the case as well as atom, and, giving due limits to the meaning of the word, it represents the state of belief in the mind of every chemist, whilst no fact whatever bears directly against it.

At the same time I do not mean to advocate the atom with the physical constitution given to it by Dalton, as well as by Newton and the ancients, not being able to see it possible;

but this is not a place for my own views: I have referred to what may be called the practical atom, or the smallest amount that unites.

Is then chemistry scientifically disposed of by this theory? as well might we say that Newton exhausted the heavens of its knowledge. Year after year will furnish us with marvellous truths, nor can we believe that centuries or millenaries will exhaust God's wisdom in the earth. Already has Davy's aluminum, a brittle useless powder not quite pure, turned a beautiful metal, and the slippery mud of a clayey soil been assimilated to "shining silver." The atomic theory may be further analyzed, and under its simple laws may be found another which will not only include all we now have, but a host of others still unsuspected; the time may even come when a new chemistry will be revealed to us, a world under our present elements, when every element will be convulsed and shaken into fragments, by powers which nature will put into our hands; but even that does not destroy the laws of the present. Even when that scientific convulsion comes, we can scarcely doubt that the elements will break up, well proportioned and according to regular laws, if they break into fragments at all. But this stratum of our knowledge cannot be annihilated by any under stratum; what we have found is true, whatever higher truths may overpower us with their splendour. When these truths come let us receive them openly and willingly, giving them encouragement instead of envious repulsion, knowing, in fact, that they must come, and rather let us make an occasional mistake in harbouring a mere mortal, than lose the opportunity of an angel for a guest. There are incredulous fools who have made the world's throbbing heart a blank to them, lest they should perhance at times be cheated. There have been madmen who have refused to eat, lest they should be poisoned.

These reflections naturally arise on considering the manner in which Dalton's discovery was reviewed. It cannot be denied, however, that the time being nearly ripe was the cause of its ultimate success. To discover out of place and out of time is a great misfortune. To be before the age in knowledge is to a man a curse, and to a generation no advantage; at least it would seem so, although there may be a value even for this occasional misplacement not to be lost sight of in estimating man's progress towards a higher civilization. But in such cases the individual is not appreciated, the generation does not hear him, his years pass by in misery, and his attempts to teach are a failure. That knowledge is best rewarded which is a fit evolution of the age, and which can be at once put to use for practical purposes, or for mental cultivation. It behoves us then to be respectful towards new opinions, and tender towards crotchets, lest we may be laughing, as so many have done before us, at beautiful truths. The rudiments of truth are no more beautiful to us than the roots of flowers, until we study them thoroughly; by forgetting this men are made victims who ought to be revered.

CHAPTER XII.

LATER LIFE OF DALTON.

ALTHOUGH the atomic theory must ever be considered as Dalton's great discovery, we find that he obtained it at the end of a series of investigations of themselves sufficient to have made him a conspicuous character. These earlier labours had great influence in advancing physical knowledge, and first brought him into repute. We find him so early as 1804 lecturing in the Royal Institution, when his theory was scarcely known but to himself, and when it was not expected to form part of a lecture. Fame was now beginning to hover round him, and he was not insensible of the change. But to persons of his habits, fame is less welcome in person than by letter, as we may say. They are less fitted for receiving the compliments of the world addressed to them in their presence, than for receiving those more impressive results,—great respect and deference; and, above all, the pleasure of seeing the influence they exercise upon society. Dalton was evidently conscious of the position to which he was entitled in the scientific world, although equally conscious that he was out of place in those brilliant assemblies in which scientific men in capitals occasionally mingle. He seems to write as if it were rather curious, although true, that he, too, was one of them, that the hope and struggle of his life to attain a position in science had been realized, but to feel at the same time that his life was not there, but as before, in his laboratory.

It not unfrequently happened, after this period, that he was engaged in controversy; and we find him there acting with great ease and deliberation, always without fear. Like most scientific men, he was destined to modify or to contradict some of his earlier conclusions, having found that he formed laws too hastily. At page 9 of his "New System," he says, "Sometime ago it occurred to me as probable, that water and mercury, notwithstanding their apparent diversity, actually expand by the same law, and that the quantity of expansion is as the square of the temperature from their respective freezing points. Water very nearly accords with this law according to the present scale of temperature, and the little deviation observable is exactly of the sort that ought to exist, from the known error of the equal division of the mercurial scale. By prosecuting this inquiry, I found that the mercurial and water scales divided according to the principles just mentioned, would perfectly accord as far as they were comparable; and that the law will probably extend to all other pure liquids; but not to heterogeneous compounds, as liquid solutions of salts."

In Vol. II.* he says, "The great deviation of the scales between the temperatures of freezing water and freezing mercury is sufficient to show, as Dulong and Petit have observed, that their coincidence is only partial:" and so acknowledges his error.

In the first volume † he also said, that "the force of steam from pure liquids, as water, ether, &c., constitutes a geometrical progression to increments of temperature in arithmetical progression." Also, "that the expansion of permanent elastic fluids is in geometrical progression to equal increments of temperature." Again,‡ "that the force of steam in contact with water, increases accurately in geometrical progression to equal increments of temperature, provided these increments are measured by a thermometer of water or mercury." This seemed an ingenious group of laws, and although he gave them up, yielding to the results of Dulong, he published them unaltered in the edition or reprint of 1842. But that

* Page 289. † Page 13. ‡ Page 11.

was his custom, even when his opinion changed; it was more with a view of obtaining copies of his own works, than with the view of continuing any law proved to be incorrect, as appears from the preface to his Meteorology.

In his "New System," Part 1st, he treats of heat and the constitution of bodies; in Part 2nd, which was published in 1810, he treats of the chemical elements. In these volumes it is remarkable how thoroughly every idea has been revolved in his own mind, and become his own, before he has ventured to write it. Every chapter shows a strict independent thought, but, on the other hand, the book is wanting in the results of others, and could never consequently be a complete system. He endeavours to construct the whole science himself, more than could be accomplished by any man. The book is written in a more attractive manner than systems of chemistry now assume; and there is a constant discussion of questions which give an insight into the state of knowledge of the time and the tendency of chemistry. Still the arrangement was not well adapted for the young student, although the work was a great fund of thought for the advanced man of science. Even now few will be able to read it without advantage.

Part 2nd is principally taken up in determining the composition of bodies and their atomic weights. In the appendix he enters into discussion with Gay Lussac. He there says, Gay Lussac's "opinion is founded upon a hypothesis that all elastic fluids combine in equal measures, or in measures that have some simple relation one to another, as 1 to 2, 1 to 3, 2 to 3, &c.; in fact, his notion of measures is analogous to mine of atoms; and if it could be proved that all elastic fluids have the same number of atoms in the same volume, or numbers that are as 1, 2, 3, &c., the two hypotheses would be the same, except that mine is universal, and his applies only to elastic fluids. Gay Lussac could not see that a similar hypothesis had been entertained by me and abandoned as untenable." In this he refers to the following in p. 188 of

Part 1st. "At the time I formed the theory of mixed gases I had a confused idea, as many have, I suppose, at this time, that the particles of elastic fluids are all of the same size; that a given volume of oxygenous gas contains just as many particles as the same volume of hydrogenous." But he arrived at the conclusion, "That every species of pure elastic fluid has the particles globular, and all of a size; but that no two species agree in the size of their particles, the pressure and temperature being the same." Then he concludes, in Part 2nd, "The truth is, I believe, that gases do not unite in equal or exact measures in any one instance; when they appear to do so, it is owing to the inaccuracy of our experiments. In no case, perhaps, is there a nearer approach to mathematical exactness, than in that of 1 measure of oxygen to 2 of hydrogen; but here the most exact experiments I have ever made, gave 1.97 hydrogen to 1 oxygen."

This discussion brings out prominently some of the points of Dalton's character. He objected to the idea of bulk being taken as a combining proportion; it was his great object to show the importance of weight and the completeness with which it answered every purpose. He conceived the combination by bulk as accidental, evidently because he had not examined the relations of the subject with sufficient care, and probably with some aversion, as his own discovery seemed in question. He is strict in the examination of the analyses of others, and seeks mathematical precision, when he could so very easily, in his own researches, overleap a few per cent. Yet no one would have been, on theoretical grounds, so likely to arrive at Gay Lussac's law as Dalton himself, as he paid most attention to the bulk of atoms, giving the relative diameters of particles of the different gases after giving the relative specific and atomic gravity. He did not always obtain an analysis so correct as that mentioned above, he continually, and to the last, insisted on the atomic weight of oxygen being 7; and he gives the specific gravity as 14 times that of hydrogen, and as he gave the specific gravity of oxygen at 1.127, he was compelled to raise very high that of hydrogen, which he made .0805. We should have supposed Dalton to be, of all men, the most fitted for seeing exactly the position which Gay Lussac's law would take, and for extending our view of it; but it was not so. His mind had probably become too much engrossed with his own views of the case; and the belief that the whole subject had been attained, came in at last to shorten his vision. It is not in science only that men show that they are not the best fitted for seeing their own position, and without this knowledge a false step is not easily prevented.

His memoirs, after this period, were generally on less important subjects. We see in them the same quality of originality, the same inclination to strike at the root of a subject, but it is done with less power; the result is rather an idea which he leaves to be worked out. Some of them are hurried, some are careless, some are unfinished. Had it lain within the scope of this memoir, I might have shown many sentences full of latent beauty, which have since budded and blossomed.

Of these the following is not the least remarkable, although it had gone farther than a mere idea; it was long put into practice by him. As far as I know, therefore, he is to be considered as the originator of analysis by volume, which has long been practised to a great extent in the manufactories of Lancashire. In this respect chemists are still behind Dalton, and have not yet got into the habit of using all those advantages which his works have offered, although the actual knowledge on the subject has advanced far beyond the period of Dalton's latest years.

In one of these memoirs on the analysis of spring and mineral waters, we find that he gives directions for the centigrade method of testing, and Mr. John Gough Watson, his pupil, informs me that he used it constantly. He says,*

* Mem. Phil. Soc., Vol. III., new series, p. 59, 1814.

"The improvements I would propose in the use of tests are, that the exact quantities of the ingredients in each test should be previously ascertained and marked on the label of the bottle; this might easily be done in most of them in the present state of chemical science. We should then drop in certain quantities of each from a dropping tube graduated into grains till the required effect was produced; then from the quantity of the test required, the quantity of saline matter in the water might be determined without the trouble of collecting the precipitate; or if this was done, the one method might be a check upon the other." This method of testing, which promises to be of such great value in saving the time of chemists, was then clearly seen by him, although it has taken several workers in the field to bring it into use in the laboratory, chemists, like others, being difficult to move into a new train of thinking and acting. At the same time the mere advice is not enough, it is needful to show how it may be accomplished in various instances; this, Dalton did partially. He gave the right directions as a master, leaving it for a long train of workmen to carry out his ideas. Still we see clearly that he was accustomed to use the graduated dropping tube, and analyze by volume. He gives directions for taking the alkalinity of water by the use of acids, and adds, that "these acids may be considered as sufficient for tests of the quantity of lime in such waters, and nothing more is required than to mark the quantity of acid necessary to neutralize the lime." Here we see that he was accustomed to take the alkalinity of waters for the carbonate of lime.

There is certainly a change in the style of these memoirs, there is less care, there are opinions thrown out and unfinished experiments which do not directly lead to benefit, and there is a diminished desire to give the ultimate laws on which phenomena depend. The mind had evidently felt that something had been achieved, which left it leisure and gave it also a right to be heard, even when it uttered only its suspicions. In these memoirs we find prominently brought forward that intense faith in his own previous results, constantly quoting what was obtained in his own mind in preference to the results obtained by the whole world besides. This gradually led to a certain amount of egotism, and a conservative belief that all work in that direction was completely finished, so that he does not seem to look forward sufficiently to any improvement or modification.

Instead of reviewing his later writings, I shall add here a list more complete than has been yet given, although all the papers were not viewed by him as important, and some were merely given in all probability to supply an occasional want of material at the meetings of the Literary and Philosophical Society; by a perusal of the titles some idea will be given of his great fertility and diligence. These titles are taken from the books of the Society. On reading them over, one is compelled to wonder at the newness and youth of much of our modern science, and to doubt, on that account, the stability of not a few of its maxims; for we find that although it is intended to represent the thoughts of nature, it is of itself not older than "a man that shall die." After Dalton read his first paper, Mr. Robert Owen read one on March 6th, 1795, entitled "Thoughts on the connection between universal happiness and practical mechanics;" and in 1797, "On the origin of opinion with a view to the improvement of the social virtues." Mr. Owen is still alive.

LIST OF DALTON'S PAPERS.

Read before the Members of the Manchester Literary and Philosophical Society.

1. October 31st, 1794. Extraordinary Facts relating to the Vision of Colors, with Observations.

2. November 27th, 1795. On the Color of the Sky, and the relation between Solar Light and that derived from Com-

bustion; with Observations on Mr. Delaval's Theory of Colors.

3. April 7th, 1798. Essay on the Mind, its Ideas, and Affections; with an Application of Principles to explain the Economy of Language.

4. March 1st, 1799. A paper, containing Experiments and Observations, to determine whether the quantity of Rain and Dew is equal to the quantity of Water carried off by the rivers and raised by evaporation; with an Inquiry into the Origin of Springs.

5. April 12th, 1799. Experiments and Observations on the Power which Fluids possess of conducting Heat; with Reference to Count Rumford's seventh essay.

6. June 7th, 1799. On the Color of the Sky, and the relation betwixt Solar Light and that derived from Combustion; with Observations on Mr. Delaval's Theory.

7. April 18th, 1800. Experimental essays, to determine the Expansion of Gases by Heat, and the *maximum* of Steam or Aqueous Vapour, which any Gas of a given temperature can admit of; with observations on the common and improved Steam Engines.

8. June 27th, 1800. On the Heat and Cold produced by the Mechanical Condensation and Rarefaction of Air.

9. October 17th, 1800. Philological Inquiry into the Use and Signification of the Auxiliary Verbs and Participles of the English Language.

10. December 12th, 1800. Review of Dr. Herschel's Experiments on the *radiant* Heat, and the Reflectibility and Refrangibility of Light.

11. July 31st, 1801. Read Part 1st of Mr. Dalton's paper on the Constitution of Mixed Gases, &c.

12. October 2nd, 1801. Read Part 2nd of Mr. Dalton's paper on the Force of Steam, &c.

13. October 16th, 1801. Read Part 3rd of Mr. Dalton's paper on Evaporation, &c.

14. January 22nd, 1802. On the General Causes, Force, and Velocity of Winds; with remarks on the Seasons most liable to high winds.

15. October 29th, 1802. On the Proportion of the several Gases or Elastic Fluids, constituting the Atmosphere; with an Inquiry into the Circumstances which distinguish the *Chymical* and *Mechanical* Absorption of Gases by Liquids.

16. January 14th, 1803. On the Spontaneous Intercourse of different Elastic Fluids, in confined circumstances.

17. October 7th, 1803. On the Absorption of Gases by Water.

18. November 4th, 1803. On the Law of Expansion of Elastic Fluids, Liquids, and Vapours.

19. February 24th, 1804. A Review and Illustration of some Principles in Mr. Dalton's course of Lectures on Natural Philosophy, at the Royal Institution, in January, 1804.

20. August 3rd, 1804. On the Elements of Chemical Philosophy.

21. October 5th, 1804. On Heat.

22. November 30th, 1804. Review of Dr. Hope's paper "On the Contraction of Water by Heat."

23. September 2nd, 1805. Remarks on Mr. Gough's two Essays on Mixed Gases, and on Mr. Schmidt's, "On Moist Air."

24. March 7th, 1806. On Respiration and Animal Heat.

25. February 6th, 1807. On the Constitution and Properties of Sulphuric Acid.

26. October 2nd, 1807. On Heat.

27. October 16th, 1807. On the Expansion of Bodies by Heat.

28. January 22nd, 1808. On the Specific Heat of Bodies.

29. March 18th, 1808. On the Specific Heat of Gaseous Bodies.

30. December 2nd, 1808. On the Measure of Mechanical Force.

31. December 16th, 1808. On Respiration.

32. March 10th, 1809. On Evaporation.

33. April 7th, 1809. On the Compounds of Sulphur.

34. November 3rd, 1809. On Muriatic Acid.

35. December 1st, 1809. On Sulphuric Acid.

36. March 9th, 1810. On Fog.

37. November 16th, 1810. Appendix to his Remarks on Respiration and Animal Heat.

38. December 28th, 1810. On Hygrometry.

39. April 3rd, 1812. On Meteorology.

40. April 17th, 1812. Meteorology continued.

41. October 2nd, 1812. On the Oxymuriate of Lime.

42. January 8th, 1813. Experiments on Phosphoric Acid and the Phosphates.

43. March 5th, 1813. Experiments and Observations on the different Compounds of Carbonic Acid and Ammonia.

44. October 1st, 1813. On the Combinations of Gold.

45. October 15th, 1813. Continuation of the paper on the Combinations of Gold.

46. November 12th, 1813. The Combinations of Platina.

47. December 10th, 1813. On the Cause of Chemical Proportion, being remarks on a paper by Berzelius.

48. January 7th, 1814. Experiments on certain frigorific Mixtures.

49. March 18th, 1814. Remarks tending to facilitate the Analysis of Spring and Mineral Waters.

50. October 7th, 1814. On Metallic Oxides.

51. December 2nd, 1814. On Metallic Oxides. (Continued.)

52. January 27th, 1815. Critical remarks on some modern Chemical Phrases.

53. November 17th, 1815. Remarks on Saussure's Essay on the Absorption of Gases by Liquids.

54. October 4th, 1816. On the chemical compounds of Azote and Oxygen.

55. December 13th, 1816. An Appendix to the Essay on chemical compounds of Azote and Oxygen.

56. October 3rd, 1817. On Phosphurets, or the combinations of Phosphorus with Earths, Alkalies, Metals, &c.

57. November 21st, 1817. Observations on Oxides and Sulphurets.

58. November 13th, 1818. Observations on the Quantity of Rain during the last Twenty-five Years; with Remarks on the Theory of Rain.

59. December 11th, 1818. Summary of Observations on the Barometer and Thermometer, made at Manchester for the last 25 years.

60. January 8th, 1819. Experiments on the Force of the Vapour of Ether, to show the fallacy of some of Dr. Ure's Statements just published in the Philosophical Transactions.

61. April 16th, 1819. On Sulphuric Ether.

62. October 15th, 1819. On Alloys, particularly those of Copper and Zinc, and Copper and Tin.

63. November 12th, 1819. On Amalgams, and other Metallic Alloys.

64. December 10th, 1819. A Chemical Analysis of the Mineral Waters of Buxton.

65. October 6th, 1820. On Oil, and the Gases obtained from it by heat.

66. December 1st, 1820. On Alum.

67. January 26th, 1821. On Meteorology, or observations on the Weather for the years 1819 and 1820, in Manchester.

68. February 9th, 1821. Observations on Meteorology, particularly with regard to the Dew point, &c., or quantity of Vapour in the Air.

69. October 5th, 1821. Some observations on the Salts and Sulphurets of Iron.

70. November 30th, 1821. On the Effects of continued Electrification on compound and mixed Gases.

71. December 13th, 1822. On the Saline Impregnations

of the Rain which fell during the late Storm, viz., December 5th, 1822.

72. March 21st, 1823. Appendix to an Essay on Salt Rain (read December 13th, 1822), with additional observations on the succeeding Storms of Wind and Rain.

73. November 14th, 1823. On the Nature and Properties of Indigo; with directions for the valuation of different samples.

74. December 26th, 1823. On various Alloys of Tin, Zinc, Lead, Bismuth, Antimony, &c.

75. October 15th, 1824. On Associations for the Promotion of the Physical Sciences, Literature, and the Arts.

76. November 12th, 1824. An Account of some Experiments to determine the Light and Heat given out by the combustion of different Gases.

77. April 15th, 1825. Results of Meteorological Observations at Manchester, for Thirty-one Years; with Remarks upon them.

78. December 30th, 1825. On the Constitution of the Atmosphere.

79. October 6th, 1826. On the Height of the Aurora Borealis above the surface of the Earth, particularly the one seen on the 29th of March, 1826.

80. November 4th, 1826. An Appendix to a paper read on October 6th, on the height of the Aurora Borealis above the surface of the earth.

81. November 26th, 1827. An Historical Sketch of the Society's Library; with an Account of its present state.

82. December 28th, 1827. Observations, chiefly chemical, on the nature of the Rock Strata in Manchester and its vicinity.

83. October 17th, 1828. Summary of the Rain, &c., at Geneva and at the elevated station of St. Bernard, for a series of years, from the 'Bibliotheque Universelle' for March, 1828; with Observations on the same.

84. January 8th, 1830. Physiological Investigations, de-

duced from the Mechanical Effects arising from Atmospherical Pressure on the Animal Frame.

85. January 22nd, 1830. Remarks on a Statement of the Amount of Rain fallen at different places on the line of the Rochdale Canal.

86. March 5th, 1830. On the Quantity of Food taken by a person in health, compared with the Quantity of the different Secretions during the same period; with Chemical Remarks on the several Articles.

87. October 15th, 1830. Chemical Observations on certain Atomic Weights, as adopted by different Authors; with some Remarks on the Notation of Berzelius.

88. October 29th, 1830. Observations on the Causes of Colouring Matter.

89. November 23rd, 1830. Chemical Observations on certain Atomic Weights, as adopted by different Authors; with Remarks on the Notation of Berzelius.

90. January 21st, 1831. Meteorological Observations for a period of thirty-seven years; with Theoretical Remarks.

91. February 18th, 1831. On the Quantity of Oxygen in Atmospheric Air.

92. December 2nd, 1831. On the Proportion of Oxygen Gas in the Atmosphere.

93. January 13th, 1832. A Summary of Meteorological Observations, for 1831, made in Manchester and the vicinity.

94. January 11th, 1833. Dr. Dalton's Remarks on the Meteorology of the last year.

95. March 8th, 1833. Observations on the Anomalous Vision of Colours.

96. November 1st, 1833. A Description of an *imaginary* Aurora Borealis in the North of England.

97. February 7th, 1834. An Account of Meteorological Observations, at Manchester and other places, in the year 1833.

98. March 7th, 1834. Some Remarks on Clouds: their Nature, Height, &c.

99. October 17th, 1834. Observations on certain Liquids obtained from Caoutchouc by Distillation.

100. December 26th, 1834. Observations on the various accounts of the Luminous Arch or Meteor accompanying the Aurora Borealis of November 3rd, 1834.

101. February 20th, 1835. Account of Meteorological Observations, made in Manchester and other places, in 1834.

102. October 2nd, 1835. Read a paper by Dr. Dalton. (Subject not named in the Journal.)

103. February 15th, 1836. An Account of Meteorological Observations, made in Manchester and other places in 1835.

104. October 21st, 1836. Sequel to an Essay on the Constitution of the Atmosphere; read to the Society in the year 1825. Part 1.

105. November 4th, 1836. 2nd Part of a paper entitled "Sequel to an Essay on the Constitution of the Atmosphere."

106. October 2nd, 1838. On Arseniates and Phosphates.

107. February 5th, 1839. Some Account of Meteorological Observations, made in Manchester, in the years 1836-37-38.

108. October 1st, 1839. On the Ammoniaco-Magnesian Phosphate, as it was formerly called; or the Tribasic Phosphates of Magnesia and Ammonia, as Professor Graham has called it. And on the Phosphate of Soda and Ammonia, or Microscopic Salt, as it was formerly called; and now Tribasic Phosphate of Soda and Ammonia and Water, of Professor Graham.

109. March 31st, 1840. On the Quantity of Acids, Bases, and Water in the different varieties of Salts; with a New Method of Measuring the Water of Crystallization.

110. April 28th, 1840. Some Account of Meteorological Observations, made in Manchester, in the year 1839.

111. October 6th, 1840. Continuation of a paper On the Quantity of Acids, Bases, and Water in the different varieties of Salts.

112. January 12th, 1841. Meteorological Observations, made in Manchester and the neighbourhood, during the year 1840, or previously.

113. March 9th, 1841. On a New and Easy Method of Analyzing Sugar.

114. October 5th, 1841. On the Citric Acid, the Oxalic Acid, the Acetic Acid, and the Tartaric Acid.

115. January 10th, 1843. Meteorological Observations, at Manchester, made in the year 1842.

116. April 16th, 1844. On the Fall of Rain, &c., &c., in Manchester, during a period of 50 years.

Some of these were embodied in other works or printed elsewhere.

In Nicholson's Journal.

New Theory of the Constitution of Mixed Gases elucidated. Vol. III., p. 267. November 18th, 1802.

Letter from Mr. Dalton, containing Observations concerning the Determination of the Zero of Heat, the Thermometrical Gradation, and the Law by which dense or non-elastic Fluids expand by Heat. Vol. V., p. 34. April 20th, 1803.

Correction of a mistake in Dr. Kirwan's Essay on the State of Vapour in the Atmosphere. Vol. VI., p. 118. August 22nd, 1803.

On the supposed Chemical Affinity of the Elements of Common Air; with Remarks on Dr. Thomson's Observations on that subject. Vol. VIII., p. 145. June 16th, 1804.

Observations on Mr. Gough's Strictures on the Doctrine of Mixed Gases, &c. Vol. IX., p. 89. September 8th, 1804.

Facts tending to Decide the Question at what Point of Temperature Water possesses the greatest Density. Vol. X., p. 93. January 10th, 1804.

Extract of a Letter from Mr. J. Dalton: On a remarkable Aurora Borealis. Vol. X., p. 303. March 12th, 1805.

Remarks on Count Rumford's experiments relating to the maximum density of water. Vol. XII., p. 28. August 17th, 1805.

Investigation of the Temperature at which Water is of the greatest Density, from the Experiments of Dr. Hope, on the Contraction of Water by Heat at low temperatures. Vol. XIII., p. 377. April 14th, 1806. And Vol. XIV., p. 128. May 12th, 1806.

Inquiries concerning the signification of the word particle, as used by modern chemical writers, as well as concerning some other terms and phrases. Vol. XXVIII., p. 81. December 19th, 1811.

Remarks on Potassium, Sodium, &c. Vol. XXIX., p. 129. May 11th, 1811.

Observations on Dr. Bostock's Review of the Atomic Principles of Chemistry. Vol. XXIX., p. 143. May 15th, 1811.

In Thomson's "Annals of Philosophy."

Further Observations and Experiments on the Combinations of Oxymuriatic Acid with Lime. Vol. II., p. 6. 1813.

Remarks on the Essay of Dr. Berzelius, on the Cause of Chemical Proportion. December 24th, 1813. Vol. III., p. 174. 1814.

Vindication of the Theory of the Absorption of Gases by Water, against the conclusions of Saussure. Vol. VII., p. 215. 1816.

On the Chemical Compounds of Azote and Oxygen, and on Ammonia. Vols. IX., p. 186, and X., p. 38 and 83. 1817.

On Phosphuretted Hydrogen. Vol. XI., p. 7. 1818.

On the Combustion of Alcohol, by the lamp without flame. Vol. XII, p. 245. 1818.

On the Vis Viva. Vol. XII., p. 444. 1818.

In Phillips's "Annals of Philosophy."

On the Analysis of Atmospheric Air by Hydrogen. Vol. X. N. S.

In the "Philosophical Transactions."

On the Constitution of the Atmosphere. 1826.

On the Height of the Aurora Borealis. 1828.

Sequel to an Essay on the Constitution of the Atmosphere; with some Account of the Sulphurets of Lime. 1837.

In the "Annales de Chimie."

Sur l' Hydrogène Phosphuré. (Extract of a Letter addressed to the Royal Academy of Sciences.) Vol. VII. 1817.

In a Separate Form. 1840.

Essay on the Phosphates and Arseniates.

On Microcosmic Salt.

On the Mixture of Sulphate of Magnesia and the Biphosphate of Soda.

Essay on the Quantity of Acids, Bases, and Water in the different varieties of Salts; with a New Method of Measuring the Water of Crystallization as well as the Acids and Bases.

On a New and Easy Method of Analyzing Sugar.

His "Meteorology," 1793.

"Grammar," 1801.

"New System of Chemistry," Part 1, 1808—Part 2, 1810; Vol. II., Part 1, 1827.

A new edition of Vol. I., Part 1, appeared in 1842, and a new edition of his Meteorology in 1834.

His many letters to periodicals in his youth need not be specially enumerated.

These labours did not pass unnoticed in this country, and still less, I may say, in foreign countries. In 1816, the French Academy elected him a Corresponding Member, although not yet a Member of the Royal Society of his own country.

Productions so numerous as these will account for all the years that Dalton had to spend. We find him still gaining his living as professional chemist, lecturer, and teacher of chemistry and mathematics; sometimes giving evidence on subjects connected with the arts, at trials in various courts; sometimes answering the inquiries of manufacturers. His uncertain position in this respect, suggested to Sir H. Davy that he might probably find it agreeable to undertake the scientific department of an expedition, fitted out by the Admiralty, for investigation in the Polar regions, under Sir John Ross. In 1818, Davy wrote to inquire if he would undertake it, but Dalton excused himself, partly on account of the hardships connected with a journey which, to one accustomed to a sedentary life, would be found too severe; and partly because he did not wish to interrupt his chemical inquiries. It is almost to be regretted that he did not undertake some such expedition. A new direction would have been given to his investigations; and a mind which had almost expended itself on its own field of view, would have had an ample new field on which new crops would, in all probability, have flourished abundantly.

About the same time, Mr. Strutt, of Derby, offered him a laboratory and a home at his house, with a salary of four hundred pounds a-year, and perfect freedom to spend his time in any manner he might consider agreeable. This he declined, on account of the loss of independence which he considered it would necessarily involve, although the offer was made on such terms as to free him from all duties whatever. This information I received from Alderman Shuttleworth, of Manchester, through whom the offer was made. But no professorship was offered him, none of the rich endowments of the country benefited Dalton. With a strange apathy these are seldom offered but to those who propose themselves;

there is no seeking out of talent, on which to confer opportunities of usefulness; and reward, which is an object less desired by men of elevated minds, comes only after the clamorous entreaties of friends.

The Royal Society had paid no attention to him, because constituted like many of the other public bodies of the country, receiving its strength by the vigour of individuals moving towards it, and so standing more in the condition of a reservoir than of a fountain. Davy proposed, in 1810, to nominate him, but he declined; probably the expense was the hindrance. There certainly are cases in which such a hindrance should not be permitted. In 1822, he was proposed and elected without his consent being asked, lest a true *nolo episcopari* should have again been uttered. It was in the summer of the same year that he visited Paris, and there had that distinguished reception and entertainment among its scientific men, that he might have looked for here in vain. His reception in Paris pleased him much; he returned much refreshed and invigorated in mind; he formed high opinions of many of the celebrated characters he met there, and the visit was spoken of with pleasure to the end of his life.

If the customs or laws of the Royal Society prevented them from early electing him, they took the first opportunity in their power of presenting him with the gold medal. In the discourse* of Sir H. Davy, on that occasion, we have his matured opinion on Dalton's position and discoveries. He there says that the medal was given "for the development of the chemical theory of definite proportions, usually called the atomic theory, and for his various other labours and discoveries in physical and chemical science." His speech at this time must be taken as his matured opinion on the subject of the atomic theory, whatever he may have in private said or been reported to say. It continues: "To Mr. Dalton

* Page 125, "Six discourses delivered before the Royal Society," by Sir H. Davy.

belongs the distinction of first unequivocally calling the attention of philosophers to this important subject. Finding that in certain compounds of gaseous bodies, the same elements always combined in the same proportion; and that when there was more than one combination, the quantity of the elements always had a constant relation, such as 1 to 2 or 1 to 3, or to 4, he explained this fact on the Newtonian doctrine of indivisible atoms, and contended that, the relative weight of one atom to that of any other atom being known, its proportions or weight in all its combinations might be ascertained; thus making the statics of chemistry depend upon simple questions in subtraction or multiplication, and enabling the student to deduce an immense number of facts from a few well authenticated, accurate, experimental results." He then mentions Bryan and William Higgins and Richter, saying, that "it is difficult not to allow the merits of prior conception as well as of very ingenious illustration to" W. Higgins. He expresses Richter's views very neatly, saying, "In his *New Foundations of Chemistry*, published in 1795, he has shewn that when neutro-saline bodies in general undergo mutual decomposition, there is no excess of alkali, earth, or acid; and he concludes, that these bodies are invariable in their relation to quantity, and that they may be expressed by numbers." He, however, continues to say,* "Mr. Dalton's permanent reputation will rest upon his having discovered a simple principle universally applicable to the facts of chemistry, in fixing the proportions in which bodies combine, and thus laying the foundation for future labours respecting the sublime and transcendental parts of the science of corpuscular motion. His merits in this respect resemble those of Kepler in astronomy.—Mr. Dalton has been labouring for more than a quarter of a century with the most disinterested views. With the greatest modesty and simplicity of character he has re-

* Page 129.

mained in the obscurity of the country, neither asking for approbation nor offering himself as an object of applause. He is but lately become a fellow of this society, and the only communication he has given to you is one, compared with his other works, of comparatively small interest; their (*i.e.* the council's) feeling on the subject is therefore pure. I am sure he will be gratified by this mark of your approbation of his long and painful labours. It will give a lustre to his character which it fully deserves; it will anticipate that opinion which posterity must form of his discoveries; and it may make his example more exciting to others in their search after useful knowledge and true glory."

Soon after this, in 1827, the first part of the second volume of his new system of chemistry was published, and we there see how the science would have grown had it been under the hand of Dalton alone. A vigorous hand he certainly had, but there were hundreds eagerly giving their attention to chemistry over all Europe, for the love of knowledge, of fame, and of humanity, so that it was driven impetuously forward, and under the direction of his own theory, had entirely left him behind; he was outrun by his own disciples; left alone by those who without him would not readily have moved.

In this volume he gives an account of the theory of the elementary nature of chlorine, but evidently without full belief, retracts gracefully his somewhat hazardous opinion that fluorine was a higher oxide of hydrogen than water, and uses the expression "oxide of potassium and sodium," without much disinclination. He was therefore willing to change his opinion, but in reality he had been so many years working to collect material for his book, that this volume was left behind before it was published. His dislike to much reading was found to be an injury to him, and many years of his life were thus found to be of little value to science. We must, therefore, in his public capacity view him now chiefly as one whose work was finished and who was enjoying the fruits of his labours, although at the same time he seems to have worked as constantly as ever, retaining the original habits but not displaying the original vigor. To work too long in one vein seems to be highly injurious, as the vein narrows the labour increases in greater proportion than the yield, and it is better for us to return to the main centre of the mine, leaving the distant veins to be worked out when the approaches shall have gradually become easier.

In 1830, the Academy of Sciences at Paris elected him foreign associate, in the place of Davy. Of this addition to his scientific rank in that society, he was unreservedly proud. As if his own country were constantly to be behind in his recognition, we find that in the year 1832, Oxford elected him doctor of civil laws. This title of D.C.L., given at the recommendation of Dr. Daubeny, was one which he seldom omitted after his name; the simplicity of his mind did not allow him at all to disguise the pleasure with which it was viewed.

Being now advanced in years, his friends were anxious to secure for him an age less laborious than his life had been. With small means, he had still saved some money, but too little for the support of his declining strength during the years of his probable life. We are told by Dr. Henry, that Mr. Babbage first suggested the propriety of applying to government for a pension, whilst Mr. Geo. W. Wood and Mr. Poulett Thomson were most active in obtaining it.

Dr. Henry informs us that the first answer the Lord Chancellor gave was, that he "was anxious to obtain some provision for him, but that it was attended with great difficulty." This is an expression at which we cannot be too much astonished, but there is no doubt of its truth, even if it did not sound highly probable. Were there so many superior men pressing on the treasures of the country that it was difficult to obtain a pittance for Dalton? The same defect again shows itself, there is no spontaneous movement, but every man must thrust himself or be thrust forward by his friends, and after having done his work by many struggles, another struggle must be made for the reward.

The following letter, written by Dr. William Henry, his intimate friend the celebrated chemist, was sent along with the formal application to government. Its beauty and truth demand for it a place in every memoir of Dalton.

DR. HENRY TO C. BABBAGE.

"Mr. Dalton never had, nor was ever given to expect any reward or encouragement whatsoever from government, and having been in habits of unreserved communication with him for more than thirty years, I can safely aver that it never occurred to him to seek it. He has looked for his reward to purer and nobler sources; to a love of science for its own sake; to the tranquil enjoyments derived from the exercise of his faculties, in the way most congenial to his tastes and habits, and to the occasional gleams of more lively pleasure, which have broken in upon his mind, when led to the discovery of new facts, or the deduction of important general laws. By the moderation of his wants and the habitual control over his desires, he has been preserved from worldly disappointments, and by the calmness of his temper and the liberality of his views, he has escaped those irritations that too often beset men who are over anxious for the possession of fame, and are impatient to grasp prematurely the benefits of its award. For a long series of years he bore neglect and sometimes even contumely, with the dignity of a philosopher, who though free from anything like vanity or arrogance, yet knows his own strength, estimates correctly his own achievements, and leaves to the world, generally although sometimes slowly just, the final adjudication of his fame. Among the numerous honours that have since been conferred on him by the best judges of scientific merit in this and other countries, not one has been sought by him. They have been without exception spontaneous offerings, prompted by a warm and generous approbation of his philosophical labours, and by the desire to cheer him onward in the same prosperous career. Deeply as he has felt these distinctions, they have never carried him beyond that sober and well regulated love of reputation, which exists in the purest minds, and is one of the noblest principles of action.

"In perfect consistency with Mr. Dalton's intellectual qualities are the moral features of his character, the disinterestedness, the independence, the truthfulness, and the integrity which through life have uniformly marked his conduct towards others. He has been taxed with plagiarism, but never was a charge more completely unfounded. Not only is he incapable of encroaching on the just rights of others, but even of taking tacitly to himself applause to which he does not feel that he is fully entitled. Of the work from which he is accused of having borrowed the outline of his atomic theory he had never once heard, until many years after the publication of his opinion on that subject. Nor is this at all extraordinary when it is considered that men like Mr. Dalton, of original and creative minds, trust rather to their own powers of research than to reading; and in the knowledge of the history of science are often surpassed by very inferior persons. This general remark applies to Mr. Dalton; but he is a discoverer in the true sense of the word. He has drawn from observed phenomena new and original views—upon these views he has founded distinct conceptions of a general law of nature;—he has traced out the conformity of that law with an extensive class of facts, many of which he himself first revealed by well-devised experiments, and he has thus secured an admiration not by having broached ingenious opinions merely, but by having worked out the evidences of those opinions by labours most sagaciously and perseveringly applied. Nor is it on the atomic theory only that his reputation must rest. It has a broader basis in his beautiful and successful investigations into the subject of heat; into the relations of air and moisture to each other; and into a variety of other topics intimately connected with the stability and advancement of chemical philosophy. I therefore agree with you that Mr. Dalton has strong claims upon the national respect and gratitude, and contend for his title to reward, even though he may not have accomplished anything bearing strictly upon the improvement of those arts and manufactures, which are the chief sources of our national wealth. For let it be remembered, that every new truth in *science* has a natural and necessary tendency to furnish, if not immediately, yet at some future time, valuable rules in *art*. Nothing is more common than that a general principle, when first developed, may admit of no obvious practical use; but that a few subsequent discoveries, made perhaps at a small expense of genius or labour, supply links which render it available first to individual, and in due course to public wealth and prosperity. Not to mention other instances, Mr. Watt derived from Dr. Black's discovery of latent heat, a guiding light to the noblest invention that has ever been placed in the hands of man, for giving him control over the physical world, and even for advancing his progress in moral and intellectual cultivation. The discovery of chlorine also in the laboratory of a retired chemist, brought forth no practical results for several years, but when found by a subsequent philosopher to quicken the whitening of unbleached cotton and linen goods, it was immediately applied by practical men to the art of bleaching, and no one can now calculate its immense influence in giving rapid circulation to the capital employed in the cotton and linen manufactures. Among the abstract truths unfolded by Mr. Dalton, it would not be difficult to point out the germs of future improvements in the practical arts generally, germs which now lie dormant in the shape of purely scientific propositions.

"But were it otherwise it would surely be unworthy of a

great nation to be governed in rewarding or encouraging genius by the narrow principle of a strict barter of advantages. With respect to great poets and great historians, no such parsimony has ever been exercised. They have been rewarded, and justly, for the contributions they have cast into the treasury of our *purely intellectual* wealth. And do not justice and policy equally demand that a philosopher of the very highest rank, one who has limited his worldly views to little more than the supply of his natural wants, and has devoted for more than forty years the energies of his powerful mind to enlarging the dominions of science, should be cherished and honoured by that country which receives by reflection the lustre of his well-earned fame? The most rigid advocate of retrenchment and economy cannot surely object to the moderate provision, which shall exempt such a man in his old age from the irksome drudgery of elementary teaching, and shall give him leisure to devote his yet vigorous faculties to reviewing, correcting, applying, and extending what he has already in great part accomplished. In one instance of recent date, a philosopher who has eminently distinguished himself in purely abstract science, has received the merited reward of a pension for life. It is most desirable then that the British government, by extending its justice to another not less illustrious, should be spared the deep reproach which otherwise assuredly awaits it, of having treated with coldness and neglect one who has contributed so much to raise his country high among intellectual nations, and to exalt the philosophical glory of the age."

The application met with success, and at the meeting of the British Association in Cambridge, it was announced by Professor Sedgwick, that the king had granted a pension of £150 to Dalton. This announcement, in the beautiful language of that eloquent man of science, has been frequently quoted, and is well known.

In 1836 the pension was increased to £300, but two years previously his brother Jonathan died, leaving for him as heir the paternal estate, which now made him comparatively wealthy. He, however, according to his strength, still continued working. So late as 1840 he published four essays, with the title "On the Phosphates and Arseniates; Microcosmic Salt; Acids, Bases, and Water; and a New and Easy Method of Analysing Sugar."

Here we have another instance of his old method of striking roughly in a new direction, and deciding at once on the whole district, little caring who was to come after to examine. He says, p. 10, "The new method of ascertaining the quantity of water in the salts is now to be discussed. I have a bottle with a stopper which just contains 572 grains of pure water, when the stopper is put on and wiped clean and dry, at the temperature of 60° Fahr. A graduated tube or jar is necessary, of 5in. or 6in. long and one quarter of an inch in diameter, to measure exactly to a grain of water. A platina wire is appended to the neck of the bottle, so as to be weighed more conveniently. An ounce, more or less, is to be weighed of any salt; it is then to be put into the bottle, capable of containing 572 grains of pure water (the water having been carefully tansferred into another glass vessel of more ample dimensions), and the salt dissolved and carefully transferred and weighed in the 572 bottle again, and the spare liquor, if any, is to be put into the narrow graduated tube.

"We have then 572 of pure water, + the pure water of the salt + the solid (or liquid water of the salt whatever it may be), all together in a liquid form, in the bottle and the narrow tube. I was greatly surprised at the results. If the salt was anhydrous, it would all go into the bottle exactly filling it to a grain; showing that the salt enters the *pores* of the water.

"If the salt contained water, the quantity of water was

measured by the narrow tube in all cases whatever, showing that the solid matter had in reality entered the pores of the water."

This principle he applied to the analysis of sugar, showing that its bulk in solution was equal to the amount of oxygen and hydrogen combined as water, the carbon not occupying any room. This rule Dalton considered as absolutely and universally true. He called it "the greatest discovery next to the atomic theory." This idea in the hands of Messrs. Playfair and Joule has had a fertile expansion, although Dalton's mode of expressing the law has been limited to certain classes of salts.* Had time and strength been given him, he would no doubt, after this commencement, have laboured well in the field of "atomic volume."

It is well when men become aware of the failure of their powers, and are willing to give up their places to those whose minds are in full vigor. The essay on the phosphates and arseniates affords, on page 12, a melancholy instance of the fate of those who overrate their strength. This sentence occurs in the form of an epitaph.—"I sent the account of the phosphates and arseniates to the Royal Society, for their insertion in the transactions. They were rejected. Cavendish, Davy, Wollaston, and Gilbert are no more." It sounds like an epitaph on himself, and the volume tells still more plainly that he had not followed the increased exactness required in science. Nevertheless, in one respect, the last pamphlet is a model of himself, the rapid, hasty work, the carelessness of the labours of others, and the new field struck out in his remarks on sugar. In one point, however, it fails; in his early life he did not work otherwise than on the most advanced ideas; amongst the phosphates and arseniates he had receded.

The position which Dalton had attained seemed to demand

* See "Memoirs of the Chemical Society." Vol. II., &c.

some public demonstration of honor in the town he had so long adorned, and his declining years suggested a permanent memorial. In 1834 his friends decided on having a marble statue, which should present a correct likeness, and for this purpose Chantrey was selected as the most suitable sculptor. Chantrey seems to have entered on the task with pleasure, and he has done it well. This statue is in the entrance hall of the Royal Institution, in Manchester; the trustees having charge of it on condition that no one shall be refused permission to look at it. Dr. Henry says the likeness is more ideal than the reality, a refinement being given to the countenance which did not exist in the bust which Chantrey first took and used as a model when engaged on the full figure. Dr. Henry's intimate knowledge of Dalton must prevent any one from entertaining a very different opinion, but a daguerreotype profile now before me taken from life,* shews not only the marked features of the thinker, which no one has denied as they were striking, but that peculiar refinement which gives the idea of the student and the gentleman. This small photograph on a silver plate, is exactly similar to the head so beautifully engraved by Stephenson, I suppose indeed that it served as the copy; every expression is the same, and every fold of the abundant white hair, nor can I see that the engraver has increased the refinement, although he has probably somewhat heightened the forehead.

In the same year, I believe, he was presented at court, a place that seemed scarcely to suit such a man, but he seems to have had no desire to evade any of his natural claims to honor, taking them as a necessary consequence of his work, neither too highly elated like the great majority who are honored, nor painfully retiring like Cavendish.

Being a Quaker, and not able to wear a sword, he was taken in the scarlet robes of an Oxford Doctor of Laws, and

* This belongs to Mr. John Parry, who assisted in taking it.

although it was feared that scarlet would scarcely suit one of the Society of *Friends*, Mr. Babbage, who took him, remarked that as he had a "kind of colour blindness, all red colours appeared to him of the colour of dirt." Mr. Babbage adds, "besides, I found that our friend entertained very reasonable views of such mere matters of form;" a remark perfectly true. Dalton was no bigot or formalist. The ceremony was rehearsed beforehand by his friends, and it passed over well, but not without remark on the length of time that he remained before the king, who detained him long enough to ask him several questions. Some say that Dalton, not imagining that he had to pass on without a word of conversation, had waited to be spoken to, and somewhat embarrassed his Majesty, in his desire to be civil, to find suitable questions to put to him. But Dalton had learnt his part well, and the reason of the honor that he had of staying a few seconds longer than anyone else, until people began to ask who he could be, is more likely to have been caused by the fact which Mr. Babbage mentions, that the king was informed of the unusual presentation by Lord Brougham, who was Lord Chancellor at the time, and nominally presented him.

Honors of various kinds soon became familiar to him, such as fellowships of societies and degrees from universities, of which the title of LL.D., from Edinburgh, was one. They came upon him as on one to whom they were welcome, but remained entirely external to him; his life had been complete without their aid, and it was too late for them to find a perfect sympathy either in his intellect or his habitual feeling.

It is to be regretted that Miss Johns has not preserved more about Dalton, as she had the ability and also his confidence; we might have obtained through her means a better picture of his mental and moral phases. In the year 1830, when the Johns family left the town, Dalton took a house close to their old residence and from that time lived alone. In 1832, at the age of 66, he ascended Helvellyn as firmly as ever, took the ladies an excursion in Cumberland to see his old friends, playfully introducing the two elder Miss Johns as his daughters, and their cousins, the younger ladies, as his granddaughters. But we cannot follow him on the hills; of them he never seems to have wearied, nor did he ever weary of his old friends there, nor of the study which first made him look on the peculiar aspects of nature at that spot. The place seemed to be a memory both for his intellect and his heart, and his love for the district shows how permanent in him were the feelings of both.

It was not until 1837 that Dalton felt in any decided manner the progress of age, although long before that time his energy and originality had diminished. Properly speaking he had, like many other gifted persons, only one period of great originality, occurring immediately at the conclusion of his education, so to speak. At that period it frequently happens that the mind makes choice of the materials with which it will work, and has some more or less distinct idea of the conclusion, whilst the rest of the life is directed to its elucidation or expansion.

On April 18th, 1837, he was disabled for a while by paralysis, and although he wrote some memoirs after this period he never entirely recovered. He fell suddenly to the floor after his usual morning's labor of recording the state of the barometer and thermometer. His friend Peter Clare was his constant companion, and to the end of his life acted both as secretary and friend, with an amount of reverence and affection that is seldom found. He noted down the observations when Dalton was ill, and took down at his request all the minute particulars connected with his illness and his medicines; for every illness and every dose was like a chemical experiment with Dalton.

In six weeks he had recovered, but on February 15th, 1838, he was again attacked, and was left very much enfeebled. His habits were not changed; idleness did not follow on weakness; he still made his experiment, but took, as he said, four times longer to perform it; still saw his friends, and kept up the average of his cheerfulness, although the sad feeling that his frame had decayed, was not absent from his mind. Of this he gave little indication, but when the conversation happened once to turn upon an eminent scientific man whom he admired, and had seen in France, he said, "Ah! he was a wreck then as I am now."

Mr. Ransome, once his pupil, and latterly his friend and medical attendant, informs us that from 1838 he required constant attendance, although he had no other attack until near his death. During the 1843-4 session of the Literary and Philosophical Society, he attended occasionally, where since 1817 he had been president. He had then lost his strength so much, that to walk across the two intervening streets to his own house in Faulkner-street, leaning on the arm of Mr. Clare, was a great exertion. His speech was feeble and inarticulate, so that he did not attempt to address the society. On May 20th, 1844, another slight fit occurred. Still we find him at his work, feebly, but regularly putting down his metereological observations. His earliest thoughts were on science, and they endured to the latest period of his life. On July 26th his servant observed that his hand was unusually tremulous, but, as Dr. Wilson observes, "there was the same care and corrective watchfulness manifested in the last stroke of his pen." He had written down "little rain this," and observing after a while that he had left out "day," he returned to complete it. On the morning of the 27th, about six o'clock, his attendant left him for a little, enforcing on him the propriety of not moving until his return. His feeble limbs had not even then got reconciled to perfect helplessness; he seems to have made an attempt to rise, but fell from the bed, and was found with his head on the floor, perfectly lifeless.

Mr. Harland, in the *Manchester Guardian* of the period, gives a copy of the last three lines he wrote: they will serve as an illustration of his method.

Day of Month.	Hour.	Ther. in.	Ther. out.	Barom.	Wind.	Strength.	Rain.	Evap.	Remarks.
July 26th.	8	65	75	30.03	S.W.	1			
——	13	65	73	30.10	Sy.	1			
——	21	60	71	30.18	S.W.	1	Little	Rain	

He had ceased for some months to make observations on the amount of rain and evaporation. Mr. Harland reckons his total observations at 200,000.

His life ended with science, and these few of his observations are therefore not out of place even when recording that solemn moment. So calm had been his life, that it is not surprising that in death his countenance should show a "beautiful repose," as the same writer observed in a memoir fitted for a more permanent place than a newspaper.

Many would like to know something more of Dalton's religious faith, and would expect to learn from his concluding words the hope and direction of his spirit, as if from its position at that moment they were able to calculate the angle of its divergence from earth. But that spirit had ceased to find utterance for itself, and we are compelled to look at the more solid points of a laborious lifetime. Scientific men are often far from orthodox Christianity, although sometimes living like saints, lives of purity, charity, devotion, and deep reverence. Dalton "did justly," he "loved mercy," he "walked humbly," he remembered carefully, as his will especially shews, the mercy due to "the fatherless and widows," and all our accounts speak of him as one to all appearance in an unusual degree "unspotted from the world." His profession was according to the spiritual, but inexpressive forms of the Society of Friends.

As is usually the case on the death of an eminent man, the first proof is furnished to many persons that he was once alive. It was suddenly known that a man of eminence had left us, and the greatest desire was shown to do his memory justice. Although the body of *Friends* to whom he belonged objected, the funeral was given up to be conducted by the authorities of the town. The remains, in a lead coffin enclosed in a solid oak one, were placed in the Town Hall, and for some days a constant flow of silent and gazing spectators passed through the building. Some objected to this form; but it is not easy to say what in such cases should be done. True honor can be given only by the mind and by the heart; but to honor any man publicly it is not enough that we feel it; it must be expressed. This was a solemn method of impressing it on all the forty thousand visitors, as well as on all the town and neighbourhood, who were aware of what was going on, and one probably which would leave a greater impression than a speech over the grave, heard only by a few. True, there was the explanation of his greatness wanted, but that can be given only to a few at a time, and in place of that there was the long continuation of the ceremony which, united with a full account of Dalton's life given in the *Manchester Guardian*, impressed the fact on many and enabled everyone to know why this man was so peculiarly selected for honour.

The funeral took place on the 12th of August. The train was nearly a mile in length, including most of the public bodies of the town, numerous private friends, and still more admirers on foot or in carriages. The town was occupied for a time with the burial of Dalton; the business ceased; the streets were thronged with numberless spectators; and the police of Manchester attended with a badge of mourning. The burial took place in the Ardwick Cemetery, on the south-east side of the city. The grave is enclosed by a strong rail, enclosing a space about twenty feet square, and contains a very plain and simple, but large, massive, and imposing covering of polished red granite, with the inscription in large letters, JOHN DALTON, and in smaller letters the date of his birth and death.

This tombstone, consisting of a solid granite pediment and overhanging slab, was not made till some years after Dalton's death, when a subscription was raised, amounting to £5,312, in order to carry out some of his original intentions, as well as to connect his name with some public benefaction, as a most fitting memorial. Dalton had originally set aside two thousand pounds for a "professorship of chemistry at Oxford, for the advancement of that science by lectures, in which the atomic theory as propounded by me, together with the subsequent discoveries and elucidations thereof, shall be introduced and explained." The desire to repair the losses sustained by Mr. Johns; to show a mark of respect and gratitude to Mr. Peter Clare, who had been so affectionate as a friend; and to Mr. Neild, at whose table he had been welcomed regularly for many years, seems to have caused him to alter his will, and to attend rather to persons than to institutions.

His will included the names of many who were near and dear to him and needed assistance.

With great respect for the affection shown in Dalton's will to his friends and relatives, the sum mentioned was subscribed to carry out in part his original intention. Since his death Owens College had been founded in Manchester. For this there have been provided two Dalton chemical scholarships, of fifty pounds for two years; two Dalton mathematical scholarships, of fifty pounds, also continuing two years; Dalton prizes from ten to twenty-five pounds, and a Dalton natural history prize of fifteen pounds.* A thousand pounds of the

* As advertised for 1856.

money has been devoted to obtain a bronze statue, copied from Chantrey's marble one. This has been considerably enlarged, so as to suit its position at the right hand of the centre of the Infirmary, the widest public space in Manchester, and beside the statues of other distinguished persons.

Thus although in Manchester pure science is, from peculiar circumstances, allowed only a humble station beside the practical, and very few young men are allowed to study it thoroughly, sufficient energy and enthusiasm have been found to obtain for Dalton a memorial which connects his name as well with the ornament of the city, as with the hopes of all those youths whose aspirations lead them to seek eminence or usefulness in the study of the natural sciences.

CHAPTER XIII.

DALTON'S INTELLECTUAL CHARACTER.

The question of greatest interest and importance connected with Dalton's personality relates to the character of his mind, not in a social point of view, for there we find that although the qualities were of the best material, they were not made prominent portions of his life, but intellectually in the faculty which caused him to place himself in history and connect his name with the knowledge of nature. Sir H. Davy has given him the highest character, when he said that "he was one of the most original philosophers of his time, and one of the most ingenious," and when he says that he "had none of the manners or ways of the world," and "was a very disinterested man." But in his sketch we do not see that respect with which a man having such a character ought to be treated. It is said for example, that "he had no ambition beyond that of being thought a great philosopher." Now this is a sneering expression, but, after all, is it not expressive of the whole aim of Davy's life? Still, at his noble ambition no one has ventured to sneer. Davy called it "glory," and united to his scientific discoveries fine poetic diction, but his love of nature was not so single as Dalton's, and although his sight was more delicate it was not so penetrating. There are few great men who have not had their peculiarities; when these arise from simplicity of character they have generally been considered to exalt their possessors. Davy's speculations on Dalton's course of thought are given at random. For example, he supposes him to have seen the works of the Higginses, but not Richter's, whereas the contrary was the case, the works of the former not having been heard of, but some of the latter. We are not justified in making such speculations without some foundation. This "character" referred to, written by Davy, and to be seen in Dr. Henry's life of Dalton, might probably have been modified before he gave it to the public, had he lived to do so. He certainly had a right to jot down his own speculations in private.

Dalton has been called a coarse experimenter. He taught himself and never advanced with the times, but there are many varieties of gifts, and we have not always found that the finest experimenter has been the greatest discoverer. The mind in reality makes the experiment first. Experiments are not made on things distant from our knowledge, but on those which approach nearest to it; a theory is therefore formed, arising from previous knowledge, or a question is asked without a theory, exactly at the turning point where a finger post is for a moment wanted. The mind always travels the road or by-road of theory, although wavering at the meeting of new roads. Now Dalton when he saw that the road must be in a certain direction, did not care to keep by it at every step, and so surveyed a great extent of territory. It was done with the quick decision and instinct of the hunter over wild ground. One only laments that on the first sight of new lands there was not the poet to burst out into song. It is this want of poetry, this constant plodding workmanship of the intellect, that has obtained so few admirers for Dalton, and has allowed men, whose fame might readily be got from a very few of his memoirs, to take a position in science and society, which ought to have been far inferior. Even scientific men have yielded to the feeling, and, like the world of fashion, have admired the gayest. But, after all, how few are the scientific men whose diction gives life to their discoveries. Life is scarcely apparent till after much nursing.

The mind never ceases to be strange to us, and if our pictures of men are incorrect, it often arises from our desire to comprehend thoroughly the whole. This is always difficult, probably impossible, still more so in the case of one so little demonstrative. But it is not here desired to describe that substratum of mind common to all men, but those striking features which stood above the average man, and are the true source of our interest.

The first thing that we observe in Dalton is clearness of conception; he knows what he thinks, and can define it. This is very clear through all his course, every thought is squared and finished. To this more than anything else I attribute his first idea of atoms. He was obliged to conceive of gas, and how could he do it without giving shapes to the parts? Gases could not be without parts, they expanded and contracted, and so the parts became essential to them.

The next thing to observe is directness. He went to his point rapidly. His experiments are simple, and, although rude, are exceedingly appropriate. It must, however, be remembered that although simplicity is at times beautiful, it cannot be attained in experiment so easily now as then. He prepared the way for more complex methods. Great clearness of conception often leads the mind to put down its results in form and figure, giving a mathematical character to them. It loses the poetical distance by working at the foreground, but does not forget that the foreground has a beauty and truth of its own.

The third characteristic is tenacity. His conceptions once formed seemed never to fade, or were with difficulty eradicated. This is natural to a mind with strong conceptions. Its own thoughts become its material, much more than anything said or done by others, and it prefers to reason from its own data, being those best known to it. This was remarkably seen in Dalton.

Fourthly. I would add rapidity of conclusion. This may arise in various ways. In a mind with weak conceptions or pictures of things, rapidity may be great, but can be of no value. In a tenacious mind like Dalton's, rapidity of conception is a combination of true ideas, so rapidly made that the steps are not seen clearly by the mind itself, and hence that inexplicable result which seems at least one of the ways in which genius is produced.

To understand the truth of these remarks the memoirs must be read and compared with the times, for Dalton, although he died but lately, flourished at the birth of true chemistry, and his work was done when Berzelius was only commencing.

By him laws were more easily treated than facts, and thought was easier than observation. His mind was one of those which especially sought laws, desiring to form the link between the mental and material. He seldom observed without reasoning, and he had no surplus observation: this made him of an absent disposition. He seldom reasoned without observing: this made him an experimenter. But in the movements of his mind, as of his body, there was a certain rigidity, which he himself seems to have felt, and for which others have endeavoured to find causes. One quality Dalton had to a degree almost unparalleled: the constancy with which he clung to his occupation of observing and generalizing.

His mind seems to stand before us as an intellectual tool, constantly planing, drilling, and boring with never ceasing activity, without any violent fits of haste, and without any seasons of absolute rest. As far as I know I have indicated what were the peculiar characteristics of that tool, but there is no doubt that other circumstances might have brought into activity a far different set of faculties. We see prominently in him how one portion of the mind was willing to devote itself to obscurity for the advancement of the others, how the faculty which reasoned on the "Mind its ideas and affections," and upon the method by which thought is expressed, after a very few struggles, gave way to the more active faculties devoted to physical nature.

Some have objected to such characters by saying that they result from a want of full development of the faculties, from a one-sidedness of the mind; but what man in this short life can cultivate one faculty highly without depriving the others of nutriment, or cultivate every faculty equally without stinting them all?

In Dalton we have found a mind which has shewn itself, as well in its operations as in its results, to be of the very highest scientific class.

CHAPTER XIV.

ARRANGEMENT OF THEORIES.

To those who have attended to the history of ideas, it must always seem a remarkable thing that in early times, among certain truths, the very culminating point should have been sometimes attained by one grand bound over all the difficulties of the journey, leaving undescribed all the ruggedness of the ground passed over. To those who have been born on this summit, there is the same difficulty in passing down to the plain, as there is for us to pass upwards: the same blasting of rocks and filling up of morasses. To say that the history of the struggle is lost to us does not explain all. In these three words, already quoted, "measure, weight, and number," as applied to creation, we see that men had looked long and carefully on the world, had admired its beauty, and believed that everything was arranged with scrupulous accuracy. The context shows that the fundamental idea was a moral one, and the reasons we now have were not then in existence; but in general terms there is the eminence seen on which they stood; we move back towards the plain, or try to reach the summit, measuring, weighing, and numbering, and proving the great mountain view to be true to the minutest particle of matter.

The scientific mind seeks to repeat the order of thought by which the divine mind arranged the universe; the artistic mind sees it completed, and rejoices in its beauty. The scientific mind has the work of the six days of creation, the artistic mind lives in the Sabbath of rest. The poetical mind refuses this class of limitation.

These remarks are called forth by the peculiar progress of some of the early opinions and the difficulties encountered when experiment began. We may arrange the numerous theories spoken of in the following manner, seeking the spirit or fundamental idea of each.

First we have one matter out of which all the others were made. This in early times was real substantial matter, as water and earth; afterwards a dynamical water or earth, or a natural force corresponding to them or underlying them. Then the four elements were made the origin of all things; but convertible. These theories, although under some aspects going under various heads, may be conveniently put under one and be called the *Allhylic*. (ἄλλος and ὕλη, interchangeable elements.)

The next, although allied, claims for itself a separate place. When there is recognised a universal matter, a *prima materia* from which all things are made, and which itself has no substantial qualities, but is capable of assuming all, it may be called the *prothylic* theory. (πρῶτη ὕλη, first element as it was called.)

When matter is made up of indivisible particles, the name *atomic* is already appropriately given. (α and τόμος, uncut.)

When particles are infinitely divisible, it may be useful to call this theory the *dia-tomic*. (διὰ and τομος cut through.)

When we find the original matter to be a force only, whether represented by a number, a point, a line, a geometrical figure, or a more abstract idea, this is the *dynamical* theory.

When there are neither forces nor atoms, nor distinct elements, nor universal and insuperable laws, nor a substratum of primary matter, the *mystic theory* seems an appropriate name.

In the early atomic theories, the only difference recognised in atoms is their shape. These theories *are mechanical.* Now we recognise many original differences forming elements. This is a *polyhylic* atomic theory.

No theory can entirely get rid of Dynamics, but it would only introduce confusion into the historical account, if we said more of it than the promoters themselves.

The old theories were after all exceedingly indefinite, notwithstanding the appearance of exactness. The Daltonian theory is remarkable chiefly for its idea of quantity. It defines composition and combination by quantity. It is mechanical, because it unites pieces of matter in a mason-like style; it fits every part and breaks none, but it is not merely mechanical. Force is required, and this is of a different kind with every species. It is polyhylic. It unites therefore more qualities than previous theories. There has never been any progress made in ascertaining in what essentially consists the peculiarity of the forces in each element. That remains for future inquiry. The inquiry has chiefly had relation to weight, and for that reason I have called Dalton's the quantitative atomic theory. Mr. Joule's discoveries in heat, although not purely chemical, have begun to introduce into chemistry, through physics, forces that can be exactly calculated other than weight.

I think it of importance that Dalton's theory as adopted by chemists should have a distinct adjective to express it. The term *quantitative atomic* connects itself with analysis which in every case leads us to the use of the theory, although the most convenient term for general use, by which also we do honor to the originator, is the *Daltonian* theory. It represents the mode of discovery by weighing and calculating, and the greatest fact treated of with regard to atoms, as viewed by chemists, that they are comparative quantities measurable by the balance: it represents also the state of chemistry since the theory was propounded, a wonderfully elaborate collection of orderly arrangements of bodies distinguished principally by their quantities. I have left out as unnecessary to this history such of the characters of atoms as Dalton held more as hypothetical than theoretical.

Those theories, which now appear only as histories, are not of necessity all extinct; for some of them a resurrection day will in all probability arrive, when their forms will be beautified and their powers exalted. It may be, too, that in the confusion of transition or the difficulties of progress, various inferior appearances may be looked on as the final triumph, and be hailed by the hasty and the short-sighted, although seen by the clearer judgment to be mere phantoms by the way, specimens of the mystic theory. The time may come when the old theory of the *prima materia*, which has deluded so many, may have a higher existence, and alchemy again become the "true philosophy." Over this, too, the dynamical theory may rise, grasping them all, and by giving clearer ideas of force, thrust into an inferior position both quantity and matter, and show us, with greater certainty, the true position of abstract power and of mind.

THE END.

INDEX.

DONATIONS RECEIVED SINCE LAST PUBLICATION,

WITH

A LIST OF THE DONORS.

SESSION 1855-6.

DONORS.	DONATIONS.
TRUSTEES OF RADCLIFFE OBSERVATORY, OXFORD	Radcliffe Astronomical Observations (Oxford Observatory). Vols. XIV. and XV. 1855-56.
ROYAL SOCIETY	Transactions of the Royal Society, London, Part 2, for 1851, to Part 2, for 1855, inclusive.
DITTO	Proceedings of the Royal Society, London. Vol. VII. Vol. VIII., Nos. 18, 19, 20, 21.
DITTO	List of Fellows.
DITTO	Address of the President, &c.
SMITHSONIAN INSTITUTION	Smithsonian Contributions to Knowledge. Vols. VI. and VII. 1854.
DITTO	The Seventh, Eighth, and Ninth Annual Report of the Board of Regents of the Smithsonian Institution.
DITTO	Melsheimer's Catalogue of the described Coleoptera of the United States, 1853.
DITTO	Directions for collecting, &c., Specimens of Natural History.
DITTO	The Annular Eclipse of May 26th, 1854.
THE PUBLISHER	Norton's Literary Register.
BOARD OF AGRICULTURE, OHIO, UNITED STATES	Ninth Annual Report of the Board of Agriculture of the State of Ohio, to the Governor.
AMERICAN PHILOSOPHICAL SOCIETY	Transactions of the American Philosophical Society, Vol. X., Part 3.
DITTO	Proceedings of the American Philosophical Society, Nos. 49-52.
THE AUTHOR	Philosophy of Physics. By A. BROWN.
ACADEMY OF SCIENCES, PARIS	Comptes Rendus de l'Academie des Sciences. T. 39 and 40.
DITTO	Extrait des Comptes Rendus.

DONORS.	DONATIONS.
ECOLE DES MINES	Annales des Mines T. V. and VI. T. VII., Livr. 1, 2, 3.
INSTITUTION OF MECHANICAL ENGINEERS, BIRMINGHAM	Proceedings of the Institution of Mechanical Engineers, for January, April, July, October, and December, 1855, and January and April, 1856.
SOCIETY OF ANTIQUARIES...	Archæologia, or Miscellaneous Tracts of the Society of Antiquaries, Vol. XXVI., and Proceedings, Nos. 41 and 42.
ROYAL SOCIETY OF EDINBURGH	Transactions of the Edinburgh Royal Society, Vol. XXI., Part 2, Proceedings of the Society for 1854-5.
GEOLOGICAL SOCIETY, DUBLIN	Journal of the Geological Society of Dublin, Vol. VI., Part 2.
PHYSICAL SOCIETY OF BERLIN	Fortschritte der Physik, VI., VII., VIII., Jahrgang.
DITTO	Neue Methode zur Vermeidung and Auffindung von Rechenfehlern. Von Dr. A. KRÖNIG.
SCOTTISH SOCIETY OF ARTS.	Transactions of the Royal Scottish Society of Arts, Vol. IV., Part 2.
LIVERPOOL LITERARY AND PHILOSOPHICAL SOCIETY.	Proceedings of the Liverpool Literary and Philosophical Society, Nos. 1, 2, 3, and 9.
PRINCIPAL SCOTT, OWENS COLLEGE.....................	Annual Reports and Examination Papers of Owens College, for the years 1851-56.
THE AUTHOR.................	The Bar and the Press.
DR. CLAY	Record of Obstetric Medicine, Vol. I. and II.
THE AUTHOR..................	Account of the Life of J. H. Swale. By T. T. WILKINSON, F.R.A.S.
T. T. WILKINSON............	Ladies' and Gentlemen's Diary, for 1856.
ROYAL INSTITUTION, LONDON	Notices of the Meetings of the Royal Institution, Part 5.
NEW ORLEANS ACADEMY OF SCIENCES	Constitution and Bye-laws of the New Orleans Academy of Sciences, 1854.
DITTO	Proceedings of the New Orleans Academy of Sciences, 1854.

DONORS.	DONATIONS.
HER MAJESTY'S GOVERNMENT	Memoirs of the Geological Survey of Great Britain, Vol. II., Parts 1 and 2.
DITTO	Records of the School of Mines, Vol I., Parts 1, 2, 3, and 4.
DITTO	Memoirs of the Geological Survey of Great Britain.
DITTO	Figures and Descriptions of British Organic Remains, Decades I., II., III., IV., VI , VII. and VIII.
DITTO	Memoirs of the Geological Survey of Great Britain, Mining Records. By R. HUNT, F.R.S.
DITTO	Educational Uses of Museums. By E. FORBES, Esq., F.R.S.
DITTO	Value of a Knowledge of Mineralogy. By W. W. SMYTH.
DITTO	Report of the Geology of Cornwall, Devon, and Somerset. By Sir H. DE LA BECHE.
DITTO	Figures and Descriptions of Palæozoic Fossils. By J. PHILLIPS.
HISTORIC SOCIETY OF LANCASHIRE AND CHESHIRE	Transactions of the Historic Society of Lancashire and Cheshire, Vols. I. to VII.
THE AUTHOR	Whewell's Philosophy of the Inductive Sciences, 2 vols.
UNITED STATES PATENT OFFICE	Report of the Commissioner of Patents (United States), for 1853, Part 2, Agriculture.
DITTO	Ditto, for 1854, Vol. I., Arts and Manufactures.
E. W. BINNEY, F.R.S.	On the Structure of certain Limestone Nodules. By J. D. HOOKER, M.D., and E. W. BINNEY, F.R.S.
THE AUTHOR	Introductory Lecture at the Chatham-street School of Medicine. By D. STONE, F.C.S.
MUSEUM AND LIBRARY COMMITTEE, SALFORD	Seventh Annual Report of the Executive Committee of the Salford Borough Royal Museum and Library.
ARUNDEL, SOCIETY	Catalogue of Select Examples of Ivory Carvings.
NATURFORSCHENDE GESELLSCHAFT IN DANZIG	Neueste Schriften der Naturforschenden Gesellschaft in Danzig, B. IV., 1-4. B. V., 1-3.
DITTO	Peterman's Mittheilungen ueber wichtige neue Erforschungen auf dem Gesammtgebiete der Geographie.
ROYAL AGRICULTURAL SOCIETY OF ENGLAND	Journal of the Royal Agricultural Society, Vols. I. to XVI., and Vol. XVII., Part 1.

DONORS.	DONATIONS.
Royal Academy Brussels.	Memoires de l'Academie Royale de Belgique, Tomes XXVIII. and XXIX.
Ditto	Bulletins de l'Academie Royale des Sciences, &c., de Belgique, Tomes XX., XXI., and XXII. Partie 1.
Ditto	Annuaire de l'Academie Royale des Sciences, &c., de Belgique, 1853-54-55.
Ditto	Memoires Couronnés et Memoires des Savants Etrangers Publiès par l'Academie Royale de Belgique, Tome XXVI., 1854-55.
The Author	A Lecture on the Sanitary Question. By D. Chadwick, Esq.
Ditto	A Lecture on Poor Rates and the Principle of Rating. By ditto.
Ditto	An Essay on Water Metres. By ditto.
The Author	Two Papers " On a New Series of Organic Bodies containing Metals." By Dr. Frankland, F.R.S., &c.
Royal Irish Academy	Transactions of the Royal Irish Academy, Vol. XXII., Part 5, Science.
Ditto	Ditto, ditto, Part 6, Polite Literature.
Ditto	Proceedings of the Royal Irish Academy, Vols. V., VI., Parts 1 and 2, for 1853-4-5.
Ditto	An Essay on the Hycsos Rulers.
Geological Society	The Quarterly Journal of the Geological Society, Vol. XI., Part 4, Nov. 1, 1855.
Ditto	Ditto, Vol. XII., Parts 1 and 2, Nos. 45 and 46.
The Author	An Essay on Meteorites. By R. P. Greg, Esq.
Library Committee, Manchester Corporation	Third Annual Report of the Working of the Manchester Free Library.
Commissioners of Patents	Alphabetical Index of Patents of Inventions.
Ditto	Ditto of Patents Applied for and Patents Granted.
Ditto	Reference Index of Patents of Inventions.
Ditto	Appendix to Reference Index.
Ditto	Subject Matter Index of Patents of Inventions, Parts 1 and 2.
Academie de Dijon	Memoires de l'Academie des Sciences.
Ditto	Memoires de l'Academie des Sciences Arts et Belles Lettres de Dijon, 1828-29-30-31-32-34-36-43-44-45-56-47-48-49-50-51-52-53-54-55, to Tome IV., 2nd ser. inclus.

DONORS.	DONATIONS,
ROYAL SAXON SOCIETY OF SCIENCES	Berichte d. Philolog. Histor. Classe, 1849, I. to V.
DITTO	Ditto ditto 1850, I. to IV.
DITTO	Ditto ditto 1851, I. to V.
DITTO	Ditto ditto 1852, I. to IV.
DITTO	Ditto ditto 1853, I. to V.
DITTO	Ditto ditto 1854, I. to VI.
DITTO	Ditto ditto 1855, I. to II.
DITTO	Abhand. d. Math. Phys. Classe I. and II., B.
DITTO	Drobisch, Nachträge uëber Musikal. Tonverhältnisse.
DITTO	Berichte d. Math. Phys., Classe 1849—1855.
DITTO	Abhand. d. Philolog. Histor., Classe 1850.
DITTO	Roscher zur Geschichte d. Engl. Volkswirthschaftslehre 1851, Nachträge 1852.
DITTO	Mommsen, Polemii Sylvii Laterculus, 1853.
DITTO	Ditto, Volusii Maeciani Distributio, 1853.
DITTO	Droysen, Everard Windeck, 1853.
DITTO	Ditto, Kaiser Karls V., Lande, 1854, and V. Wietersheim Gedächtnisslehre, 1854.
ACADEMY OF SCIENCES OF THE INSTITUTE OF BOLOGNA	Memorie della Academia delle Scienze dell' Istituto di Bologna, Tom I., II., III., IV., and V.
PROFESSOR SEDGWICK	On the British Palæozoic Rocks and Fossils, with Plates. By SEDGWICK and Mc.COY.
MR. ROUTLEDGE	Letter on the Operation of the Smoke Nuisance Act. By C. W. WILLIAMS, 3 copies.
ADMINISTRATION DES MINES DE RUSSIE	Compte Rendu Annuel par le Directeur de l' Observatoire Physique Central. A. T. KUPFFER, St. Petersburgh, 1854 and 1855.
THE AUTHOR	Useful Information for Engineers. By W. FAIRBAIRN, F.R.S., &c., 1856.
CHEMICAL SOCIETY, LONDON	Quarterly Journal of the Chemical Socieiy of London, Vol. VIII., No. 4, and Vol. IX., Nos. 1 and 2.
INSTITUTO, LOMBARDO	Giornale dell I. R. Istituto Lombardo di Scienze Lettere ed Arti. Bibliotheca Italiana, Nuove Serie, Vols. I., II., III., IV., V., VI., VII., 1847-55.
THE AUTHOR	Introductory Lecture on the Harmony of Theory and Practice in Mechanics. By W. C. M. RANKINE, C.E., &c., University, Glasgow.
ZOOLOGICAL BOTANICAL SOCIETY, VIENNA	Verhandlungen des Zoologisch botanischen vereins in Wien, Bd. I., III., IV., and V., 1, 2. 3.

DONORS.	DONATIONS.
IMPERIAL GEOLOGICAL INSTITUTION, VIENNA ...	Jahrbuch der K. K. Geologischen Reichsanstalt, Vol. I.
THE AUTHOR...................	Recherches sur la Vegetation. Par M. VILLE.
LINNEAN SOCIETY..........	Journal of the Proceedings of the Linnean Society, Vol. I., Nos. 1, 2.
THE AUTHOR...................	Narrative, &c., of the Association for obtaining a Uniform Decimal System. By JAMES YATES, Esq.
SOCIETY OF NATURAL SCIENCES, CHERBOURG ..	Mémoires de la Société des Sciences Naturelles de Cherbourg, Tome I. and II., 1853-4.
DITTO	Notice a' l' anciennes fabriques de draps, de Cherbourg. Par A. LE JOLIS, de Cherbourg, 1854.
DITTO	Notice sur l' origine et l' etablissement de la Foire St. Clair de Querqueville. Par M. A. LE JOLIS, 1855.
DITTO	Observations sur les Ulex des environs de Cherbourg. Par A. LE JOLIS, 1853.
DITTO	Examen des espéces confondues sous le nom de *laminaria digitata.* Auct. par M. A. LE JOLIS, 1855.
ROYAL ASIATIC SOCIETY...	Transactions of the Royal Asiatic Society, Vol. III., Parts 1, 2, and 3, with Appendix.
DITTO	Journal of the Royal Asiatic Society, Nos. 13 to 18, and Vols. X. to XVI., Part 2 inclusive.
DITTO	Descriptive Catalogue of the Historical Manuscripts in the Arabic and Persian Languages, in the Library of the Royal Asiatic Society.
DITTO	Essay on the Architecture of the Hindoos.
DITTO	Maps, Nos. 1 and 2, Vestiges of Assyria, and Map of Central Assyria. By FELIX JONES.
THE AUTHOR...................	Picture of a Manufacturing District. By E. POTTER, F.R.S.
DITTO	On Rubian and its Products of Decomposition, Part 3. By E. SCHUNCK, Ph.D., F.R.S., &c.
ZOOLOGICAL SOCIETY, LONDON	Proceedings of the Zoological Society of London, 5 vols., 1850-55.
THE AUTHOR...................	A Lecture on Strikes. By J. A. NICHOLLS, F.R.A.S.
DITTO	A Natural History of Ferns, Parts 1 to 20 inclus. By A. J. LOWE, F.R.A.S. and E., Nottingham.
DITTO	The Present Crisis in Administrative Reform. By J. P. GASSIOT.
LINNEAN SOCIETY OF NORMANDY	Mémoires de la Société Linnéene du Calvados, 1824-25. 2 vols., and Atlas.

DONORS.	DONATIONS.
LINNEAN SOCIETY OF NORMANDY	Memoires de la Société Linnéene de Normandie, 1834 to 1853. 4 vols., ditto.
DITTO	Seance Publique, 1833 and 1836.
THE AUTHOR	Wreck of the Rothsay Castle. By J. ADSHEAD, Esq.
DITTO	Prisons and Prisoners. By ditto.
DITTO	Distress in Manchester, 1840-42. By ditto.
DITTO	Our Present Gaol System. By ditto.
DITTO	On Juvenile Criminals' Reformatories. By ditto.
IMPERIAL ACADEMY OF METZ	Mémoires de l'Académie Imperiale de Metz, 1840-46. 1848-49-50, 1853-54-55. Total, 14 vols.
SOCIETY OF ANTIQUITIES, COPENHAGEN	Mémoires de la Société des Antiquaires, Copenhague, 1845-49, Vol. I., 1852.
THE AUTHOR	Aus der Hagia Sophia. Von PAULUS CASSELL, Secretär der Erfurter Academie.
NATURAL HISTORY SOCIETY, GENEVA	Mémoires de la Société de Physique et d'Histoire Naturelle de Genéve, Tome XIV., Partie 1, 1855.
NATURAL HISTORY SOCIETY, BONN	Verhandlungen des Naturhistorischen Vereins der Preussischen Rheinlande und Westphalens, 1855, 1856.
DITTO	Sitzungsberichte und Correspondenz Blatt, 1855, 1856.
ROYAL CORNWALL POLYTECHNIC SOCIETY	Twenty-third Annual Report of the Royal Cornwall Polytechnic Society.
LEOPOLD CAROLINE ACADEMY OF NATURALISTS, BRESLAU	Acta Academiae Naturae Curiosorum, Vol. XXV., Part 1, 1855.
ROYAL ACADEMY OF SCIENCES AND BELLES LETTRES, NAPLES	Rendiconto della SocietâReale Borbonica Academia delle Scienze, Anno IV., 1855.
DITTO	Eruzioni Vesuviane, del 1850, E 1855.
BAVARIAN ACADEMY OF SCIENCES	Abhandlungen d. Akad. d. Wissenschaften histor. u. Philos. Inhalts B. I-X., 1763-1776.
DITTO	Neue Philos. Abhandlungen, I-VII., 1778-97.
DITTO	Physik. Abhand , 1-2, 1803-6.
DITTO	Berichte üeber d. Arbeiten d. Math. Phys. Classe, 1-4, 1807-11.
DITTO	Gelehrte Anzeigen. B. 38-40, 41.

DONORS.	DONATIONS.
BAVARIAN ACADEMY OF SCIENCES.	Bulletin pro 1847-53.
DITTO	Denkschriften, 1808-24.
DITTO	Abhand. Math. Phys., 1829-1854.
DITTO	Pettenkofer. Chemie.
DITTO	Buchner. Antheil.
DITTO	Vogel. Chemisch. Laboratorium.
DITTO	Vogel. Chemismus.
DITTO	Kuhn. Clima.
DITTO	Hermann. Bewegung d. Bevölkerung.
DITTO	——— Gliederung. Ditto.
DITTO	Lamont. Denkrede.
DITTO	Abhandlungen der Philos. Philolog. Classe, Vol. VII., abth. 3.
DITTO	Annalen der Sternwarte, Vols. VII., VIII.
DITTO	Almanach, 1855.
IMPERIAL ACADEMY OF ROUEN	Précis Analytique de l' Académie de Rouen, 1854-55.
YORKSHIRE GEOLOGICAL SOCIETY	Report of the Proceedings of the Geological and Polytechnic Society of the West Riding of Yorkshire, 1855.
LEEDS LITERARY AND PHILOSOPHICAL SOCIETY.	Thirty-sixth Report of the Council of the Leeds Philosophical and Literary Society, 1855-56.
GEOGRAPHICAL SOCIETY	Proceedings. Anniversary Address of the Royal Geographical Society, 1856.
ROYAL INSTITUTION	Notices of the Meetings of the Members of the Royal Institution of Great Britain. Part 6. July, 1855, to July, 1856.
ROYAL SOCIETY OF EDINBURGH	Transactions of the Royal Society of Edinburgh. Vol. XXI., part 3, 1855-56.
DITTO	Proceedings of ditto session 1855-56.
ROYAL SOCIETY OF STOCKHOLM	Proceedings of the Royal Society of Stockholm, 1844 to 1855 inclusive, 12 vols.
DITTO	Transactions of ditto 1844 to 1854 inclusive, 11 vols.
ADMINISTRATION OF THE IMPERIAL MINES, RUSSIA.	Annales de l' Observatoire Physique Central de Russie, 1851-52-53. Published 1853-54-55. 5 volumes. (St. Petersburgh.)

THE COUNCIL

OF THE

Literary and Philosophical Society of Manchester.

1856-57.

President.

WILLIAM FAIRBAIRN, F.R.S., Inst. Nat. Sc., Paris Corresp.

Vice-Presidents.

JAMES PRESCOTT JOULE, F.R.S., &c.
JOSEPH CHEESEBOROUGH DYER.
THOMAS HOPKINS.
EDWARD WILLIAM BINNEY, F.R.S., F.G.S.

Secretaries.

ROBERT ANGUS SMITH, Ph.D., F.C.S.
EDWARD SCHUNCK, Ph.D., F.R.S.

Treasurer.

HENRY M. ORMEROD.

Librarian.

CHARLES FREDERIC EKMAN.

Of the Council.

RICHARD ROBERTS, M.Inst.C.E.
HENRY BOWMAN.
JAMES WOOLLEY.
PETER SPENCE.
REV. HENRY HALFORD JONES, F.R.A.S.
PROFESSOR R. C. CHRISTIE, M.A.

ALPHABETICAL LIST OF THE MEMBERS

OF

The Literary and Philosophical Society of Manchester.

1856-57.

	DATE OF ELECTION.
Ralph F. Ainsworth, M.D	April 30th, 1839
James Allan, Ph.D., M.A	January 24th, 1854
W. H. Ash	April 17th, 1849
Thomas Ashton, *Hyde*	August 11th, 1837
John Atkinson	January 27th, 1846
Joseph Adshead	April 29th, 1856
Robert Barbour	January 23rd, 1824
Henry Bernoulli Barlow	January 27th, 1852
Joseph Barratt	April 19th, 1842
Rev. R. Bassnett, A.M.	April 17th, 1849
John F. Bateman, M.Inst.C.E., F.G.S., &c.	January 21st, 1840
Thomas Bazley	January 26th, 1847
Thomas Sebastian Bazley, B.A	April 19th, 1853
William Bell	January 26th, 1847
James Bevan	January 23rd, 1844
Edward William Binney, F.R.S., F.G.S	January 25th, 1842
Alfred Binyon	January 26th, 1838
Richard Birley	April 18th, 1834
James Black, M.D., F.G.S.	April 30th, 1839
John Blackwall, F.L.S.	January 26th, 1821
Henry Bowman	October 29th, 1839
Eddowes Bowman, M.A.	January 23rd, 1855
Charles Beyer	January 24th, 1854
William Brockbank	April 17th, 1855
Edward Brooke	April 30th, 1824
W. C. Brooks, M.A.	January 23rd, 1844
Frederick Broughton	April 17th, 1855
Henry Browne, M.B.	January 27th, 1840
Laurence Buchan	November 1st, 1810

	DATE OF ELECTION.
John Burd	January 27th, 1846
Henry Bury	April 19th, 1853
William Callender, jun.	January 24th, 1854
Frederick Crace Calvert, M.R.A.T	January 26th, 1847
John Young Caw, F.S.A., F.R.S.L	April 15th, 1841
Charles E. Cawley, C.E.	January 23rd, 1855
David Chadwick	April 20th, 1852
Henry Charlewood	January 24th, 1832
Richard Copley Christie, M.A., Owens College	April 18th, 1854
Charles Clay, M.D	April 15th, 1841
Charles Cleminshaw	April 29th, 1851
Samuel Clift	April 19th, 1853
Thomas Cooke	April 12th, 1838
Edward Corbett	January 25th, 1853
Samuel Cottam	January 25th, 1853
Samuel Crompton	April 29th, 1851
James Crossley	January 22nd, 1839
Joseph S. Crowther	January 25th, 1848
Richard S. Culley, C.E	January 24th, 1854
Matthew Curtis	April 18th, 1843
John Dale	February 7th, 1854
John Benjamin Dancer, F.R.A.S	April 19th, 1842
Robert Dukinfield Darbishire	April 19th, 1853
David Reynolds Davies	January 24th, 1854
James Joseph Dean	November 15th, 1842
William L. Dickinson	Ja uary 23rd, 1855
Joseph Cheeseborough Dyer	April 24th, 1818
Frederick Nathaniel Dyer	April 30th, 1850
The Right Hon. the Earl of Ellesmere, F.R.S., &c.	April 15th, 1841
Charles Ellis	January 24, 1854
Charles Frederic Ekman	April 29th, 1856
George Fairbairn	January 25th, 1853
Thomas Fairbairn	April 30th, 1850
W. Fairbairn, F.R.S., Inst. Nat. Sc., Par. Corresp.	October 29th, 1824
W. A. Fairbairn	October 30th, 1849
David Gibson Fleming	January 25th, 1852
Richard Flint	October 31st, 1818
Benjamin Fothergill	January 23rd, 1855
H. R. Forrest	April 29th, 1856

	DATE OF ELECTION.
E. Frankland, Ph.D., F.R.S., Owens College	April 29th, 1851
Alfred Fryer	January 24th, 1854
Rev. William Gaskell, M.A.	January 21st, 1840
Samuel Giles	April 20th, 1836
John Gould	April 20th, 1847
John Graham	August 11th, 1837
Robert Hyde Greg, F.G.S.	January 24th, 1817
Robert Philips Greg	October 30th, 1849
John Clowes Grundy	January 25th, 1848
Richard Hampson	January 23rd, 1844
John Hawkshaw, F.R.S., F.G.S., and M. Inst. C.E.	January 22nd, 1839
William Charles Henry, M.D., F.R.S	October 31st, 1828
John Hetherington	January 24th, 1854
Sir Benjamin Heywood, Bart., F.R.S	January 27th, 1815
James Heywood, M.P., F.R.S., and G.S.	April 26th, 1833
James Higgin	April 29th, 1851
James Higgins	April 29th, 1845
Peter Higson	October 31st, 1848
John Hobson	January 22nd, 1839
E. Hodgkinson, F.R.S., M.R.I.A., F.G.S., &c.	January 21st, 1820
George Holcroft	January 24th, 1854
Isaac Holden	January 23rd, 1855
James Platt Holden	January 27th, 1846
Thomas Hopkins	January 18th, 1823
Henry Houldsworth	January 23rd, 1824
William Jennings Hoyle	February 7th, 1854
John Jesse, F.R.S., R.A.S., and L.S.	January 24th, 1823
Richard Johnson	April 30th, 1850
William Johnson	January 22nd, 1856
Rev. Henry Halford Jones, F.R.A.S.	January 22nd, 1856
Joseph Jordan	October 19th, 1821
B. St. J. B. Joule	April 18th, 1848
James Prescott Joule, F.R.S., &c.	January 25th, 1842
William Joynson	January 27th, 1846
Samuel Kay	January 24th, 1843
John Lawson Kennedy	January 27th, 1852
James Kershaw, jun.	January 24th, 1854
Richard Lane	April 26th, 1822

	DATE OF ELECTION.
William Langton	April 30th, 1830
Joseph Leese, jun.	April 30th, 1850
Thomas Taylor Lingard	January 22nd, 1856
Thomas Littler	January 27th, 1825
Joseph Lockett	October 29th, 1839
Benjamin Love	April 19th, 1842
Isaac Waithman Long, F.R.A.S.	January 27th, 1852
George Cliffe Lowe	January 24th, 1854
Edward Lund	April 30th, 1850
George T. Lund	January 23rd, 1855
William Tudor Mabley	October 30th, 1855
James M'Connel	October 30th, 1829
William M'Connel	April 17th, 1838
Alexander M'Dougall	April 30th, 1844
John Macfarlane	January 24th, 1823
Edward William Makinson, M.A.	October 20th, 1846
The Right Rev. the Lord Bishop of Manchester, D.D., F.R.S., F.G.S.	April 17th, 1849
Robert Manners Mann	January 27th, 1846
Thomas Mellor	January 25th, 1842
William Mellor	January 27th, 1837
John Moore, F.L.S.	January 27th, 1815
David Morris	January 23rd, 1849
Alfred Neild	January 25th, 1848
William Neild	April 26th, 1822
James Emanuel Nelson	January 27th, 1852
Thomas Henry Nevill	February 7th, 1854
John Ashton Nicholls, F.R.A.S.	January 21st, 1845
William Nicholson	January 26th, 1827
Edward Byron Noden	October 31st, 1854
Henry Mere Ormerod	April 30th, 1844
George Parr	April 30th, 1844
John Parry	April 26th, 1833
George Peel, M.Inst.C.E.	April 15th, 1841
Henry Davis Pochin	January 24th, 1854
John Ramsbottom	February 7th, 1854
Joseph Atkinson Ransome, F.R.C.S.	April 29th, 1836
Thomas Ransome	January 26th, 1847

DATE OF ELECTION.

Richard Roberts, M.Inst.C.E.. ....January 18th, 1823
Samuel Robinson.. ....January 25th, 1822
Alan Royle ....January 25th, 1842

Samuel Salt ....April 18th, 1848
Archibald Sandeman, M.A., Owens College..April 29th, 1851
Michael Satterthwaite, M.D..January 26th, 1847
Edward Schunck, Ph.D., F.R.S. ....January 25th, 1842
Alexander J. Scott, M.A., Principal, Owens College ....February 7th, 1854
Edmund Hamilton Sharp ....January 23rd, 1855
John Shuttleworth ....October 30th, 1835
Joseph Sidebotham ....April 20th, 1852
George S. Fereday Smith, M.A., F.G.S. ....January 26th, 1838
Robert Angus Smith, Ph.D., F.C.S. ....April 29th, 1845
Peter Spence..April 29th, 1851
Thomas Standring ....January 27th, 1852
Edward Stephens, M.D. ....January 24th, 1834
James Stephens..April 20th, 1847
Robert Stuart ....January 21st, 1814

John Edward Taylor ....January 22nd, 1856
David Thom..April 20th, 1852
James Thompson..April 18th, 1854
James Aspinal Turner. ....April 29th, 1836
Thomas Turner, F.R.C.S. ....April 19th, 1821

Absalom Watkin..January 24th, 1823
Joseph Whitworth ....January 22nd, 1832
Samuel Walker Williamson ....April 19th, 1853
William Crawford Williamson, F.R.S., Owens College.. ......April 29th, 1851
George Bancroft Withington ....January 21st, 1851
William Rayner Wood ....January 22nd, 1830
Alonza B. Woodcock. ....October 30th, 1855
George Woodhead ....April 21st, 1846
Edward Woods..April 30th, 1839
James Woolley..November 15th, 1842
Robert Worthington, F.R.A.S..April 28th, 1840

HONORARY MEMBERS.

Rev. William Turner, *Manchester.*

Sir Oswald Mosley, Bart., *Rolleston Hall.*

Rev. Adam Sedgwick, M.A., F.R.S., Hon. M.R.I.A., &c., *Cambridge.*

General Sir Thomas Makdougall Brisbane, Bart., F.R.S., Hon. M.R.I.A., Instit. Nat. Sc., Paris Corresp., &c., *Makerstoun, Kelso.*

Rev. William Venables Vernon Harcourt, M.A., F.R.S., Hon. M.R.I.A., F.G.S., *York.*

Rev. William Whewell, B.D., F.R.S., Hon. M.R.I.A., F.R.A.S., &c., *Cambridge.*

Sir William Hamilton, Bart., *Dublin.*

Baron Von Liebig, *München.*

Eilert Mitscherlich, *Berlin.*

Sir John Frederick William Herschel, Bart., D.C.L., F.R.S. L. & E., &c., &c., Instit. Nat. Sc., Paris Corresp.

Michael Faraday, Esq., D.C.L., Hon. Mem. R.S. Ed., Instit. Nat. Paris Socius.

George Biddell Airy, Esq., M.A., D.C.L., F.R.A.S., F.R.S., &c., &c., *Royal Observatory.*

Sir David Brewster, F.R.S. L. & E., Instit. Sc. Paris Socius, Hon. M.R.I.A., F.G.S., F.R.A.S., &c., *St. Andrew's.*

Very Rev. George Peacock, D.D., F.R.S., F.G.S., F.R.A.S., *Ely.*

Baron Alexander Von Humboldt, *Berlin.*

Peter Barlow, Esq., F.R.S., F.R.A.S., Hon. M.P.C.S., Instit. Nat. Sc., Paris Corresp., *Woolwich.*

Rev. Henry Moseley, M.A., F.R.S., *Wandsworth.*

Louis Agassiz, *Cambridge, Massachussets.*

Lieut.-Colonel Edward Sabine, R.A., F.R.S.V.P., F.R.A.S., &c.

Jean Baptiste Dumas, *Paris.*

Sir Roderick Impey Murchison, G.C. Sc. S., M.A., F.G.S., Hon. M.R.S. Ed., R.I.A., &c., &c.

Richard Owen, Esq., M.D., L.L.D., F.R.S., Hon. M.R.S. Ed., Instit. Nat. Sc., Paris Corresp., &c., &c.

John Couch Adams, Esq., F.R.S., F.R.A.S., &c., *Cambridge.*

W. J. J. Le Verrier, *Paris.*

J. R. Hind, Esq., *Regent's Park.*

Rev. Joseph Bosworth, L.L.D., F.S.A., F.R.S., &c.

Robert Rawson, Esq., *Portsmouth.*

Bennet Woodcroft, Professor, *University College.*

William Thompson, Professor, *University, Glasgow.*

Lyon Playfair, C.B., F.R.S., &c., *Department of Science and Art, Marlborough House.*

George Gabriel Stokes, M.A., &c., &c., *Cambridge.*

Rev. Thomas Penyngton Kirkman, M.A., &c., *Croft Rectory, near Warrington.*

William Hopkins, Esq., F.R.S., Pres. Geo. Soc., &c., *Cambridge.*

John Hartnup, F.R.A.S., *Observatory, Liverpool.*

General M. A. Morin, Mem. Instit. France, &c., Paris.

Rev. John James Tayler, B.A., Principal of Manchester New College, *London.*

General Poncelet, Mem. Instit. France, &c., Paris.

CORRESPONDING MEMBERS.

Rev. John Kenrick, M.A., *York.*

Benjamin Dockray, Esq., *Lancaster.*

Jonathan Otley, Esq., *Keswick.*

Rev. Robert Harley, *Brighouse, near Huddersfield.*

William Thaddeus Harris, Esq., *Cambridge, Massachussets.*

Peter Pincoffs, M.D., *Dresden.*

Henry Hough Watson, Esq., *Bolton.*

John Mercer, *Oakenshaw.*

M. Girardin, *Rouen.*

T. T. Wilkinson, F.R.A.S., &c., *Burnley.*

T. SOWLER AND SONS, PRINTERS, MANCHESTER.

Made in the USA
Monee, IL
04 November 2026

67385025R00184